Women in Engineering
Pioneers and Trailblazers

Other Titles of Interest

Becoming Leaders: A Practical Handbook for Women in Engineering, Science, and Technology, by F. Mary Williams and Carolyn J. Emerson (ASCE Press, 2008). Helps women who work hard also work smart, blending best research with practical tips. (ISBN 978-0-7844-0920-6)

Changing Our World: True Stories of Women Engineers, by Sybil E. Hatch (ASCE Press, 2006). Highlights the real-life stories of hundreds of women engineers, celebrating their contributions to every aspect of modern life. (ISBN 978-0-7844-0841-4)

Diversity by Design: Guide to Fostering Diversity in the Civil Engineering Workforce, by Sybil Hatch (ASCE, 2008). Provides hands-on suggestions to foster, improve, and maintain a diverse and thriving workforce within the civil engineering profession. (ISBN 978-0-7844-0983-1)

Engineering Legends: Great American Civil Engineers, by Richard G. Weingardt (ASCE Press, 2005). Chronicles the personal lives and professional accomplishments of 32 great U.S. civil engineers from the 1700s to the present. (ISBN 978-0-7844-0808-8)

The 21st-Century Engineer: A Proposal for Engineering Education Reform, by Patricia D. Galloway (ASCE Press, 2008). Presents a clarion call to reform the way today's engineers prepare for tomorrow. (ISBN 978-0-7844-0936-7)

Women in Engineering: Professional Life, by Margaret E. Layne (ASCE Press, 2009). Presents multiple perspectives on women in engineering, focusing on career advice, sociological studies, and agendas for action. (ISBN 978-0-7844-0991-6)

Women in Engineering

Pioneers and Trailblazers

MARGARET E. LAYNE, P.E.

Library of Congress Cataloging-in-Publication Data

Women in engineering : pioneers and trailblazers / Margaret E. Layne, [editor].
    p. cm.
  Includes bibliographical references and index.
  ISBN 978-0-7844-0980-0
  1. Women in engineering. 2. Women engineers. I. Layne, Margaret.

  TA157.W674 2009
  620.0082—dc22

                                    2009002097

Published by American Society of Civil Engineers
1801 Alexander Bell Drive
Reston, Virginia 20191
www.pubs.asce.org

Any statements expressed in these materials are those of the individual authors and do not necessarily represent the views of ASCE, which takes no responsibility for any statement made herein. No reference made in this publication to any specific method, product, process, or service constitutes or implies an endorsement, recommendation, or warranty thereof by ASCE. The materials are for general information only and do not represent a standard of ASCE, nor are they intended as a reference in purchase specifications, contracts, regulations, statutes, or any other legal document.

ASCE makes no representation or warranty of any kind, whether express or implied, concerning the accuracy, completeness, suitability, or utility of any information, apparatus, product, or process discussed in this publication, and assumes no liability therefor. This information should not be used without first securing competent advice with respect to its suitability for any general or specific application. Anyone utilizing this information assumes all liability arising from such use, including but not limited to infringement of any patent or patents.

ASCE and American Society of Civil Engineers—Registered in U.S. Patent and Trademark Office.

Photocopies and reprints. You can obtain instant permission to photocopy ASCE publications by using ASCE's online permission service (http://pubs.asce.org/permissions/requests/). Requests for 100 copies or more should be submitted to the Reprints Department, Publications Division, ASCE (address above); e-mail: permissions@asce.org. A reprint order form can be found at http://pubs.asce.org/support/reprints/.

Front cover photograph of Ellen Swallow Richards and MIT faculty courtesy of the MIT Museum. Back cover photograph of Margaret Ingels © University of Kentucky, all rights reserved, Margaret Ingels Collection, Special Collections and Digital Programs, University of Kentucky Libraries.

Contents

Preface

WHEN I ENTERED ENGINEERING SCHOOL IN 1976, FRESH out of high school, I knew that it was unusual for a girl to be an engineer. I was one of few girls in my high-school calculus and physics classes, and the only girl to sign up for an elective science class in eighth grade. But this was the 1970s, second-wave feminism was riding high following the enactment of Title IX of the Education Amendments of 1972, the law forbidding gender discrimination in educational programs receiving federal funding, women were streaming into law and medical schools in increasing numbers, and I had no doubt that more young women would be attracted into engineering as well. I couldn't imagine that more than 30 years later women would still be less than 20% of engineering students and 12% of engineering faculty, and I would leave the practice of engineering to work full time on increasing the participation and advancement of women in the profession.

Along the way I became a licensed professional engineer, worked in engineering design, and moved into management—a traditional engineering career path. I became active in professional societies and took on leadership roles. I met wonderful, accomplished women who had entered the profession before I was born and pioneered in many different fields, women who had faced active opposition and discouragement in their careers instead of the more subtle snubs and chilly climate that I learned to ignore. All the time, I wondered where all the young women were, and why they were not choosing to embrace engineering careers. More and more women were becoming doctors, lawyers, politicians, and business executives, but not engineers. Why?

As I learned about the engineering profession and about the women who found their place in it, I began to seek out more information about these women, their lives and careers, and about the engineering profession and its place in society. It was not an easy task. I found that engineering, if addressed at all, is often combined with or subsumed under the heading of science, and while women are also scarce in most scientific disciplines, women scientists still receive more attention than women engineers. I found chapters on women engineers buried here and there in books on women in science, in conference proceedings, and in government reports. My filing cabinets gradually collected these bits and pieces, so that when, in a casual hallway conversation at a professional society conference, an editor asked whether it would be feasible to publish an anthology on women engineers, I answered enthusiastically, "Yes!" In the end the collection became two volumes and I even had to make some difficult choices regarding which pieces to include and which to leave for future projects. The first volume of this collection, *Pioneers and Trailblazers*, focuses on the history of early women engineers and includes both profiles of individual women and analyses of the experiences of women engineers as a group. The second volume, *Professional Life*, presents examples of efforts to encourage more women to pursue engineering careers and to understand the social and cultural aspects of the engineering profession that impact women's experiences.

Many books about women in science have been published in the last 25 years, most notably Margaret Rossiter's two-volume *Women Scientists in America* (1982 and 1995) (which contains some information about women engineers). Ambrose et al.'s *Journeys of Women in Science and Engineering: No Universal Constants* (1997) includes some profiles of late twentieth-century women engineers, and biographies of notable women scientists are becoming more available, but still few books focus explicitly and exclusively on women engineers. Most histories of engineering that have been written rarely mention women. While women received degrees in engineering in the United States beginning in the late nineteenth century, until recently our numbers have been so small as to be virtually invisible.

Why should we care about engineers in general and women engineers in particular? Engineering is one of the largest professions in the United States, second only to teachers, with between 2.5 and 3 million practitioners, according to the Commission on Professionals in Science and Technology (2004). According to the National Science Board, "science and technology have been and will continue to be engines of U.S. economic growth and national security" (National Science Board 2003, 1). The pool of individuals from which engineers in this country have traditionally been drawn, white males, is a declining fraction of the U.S. population. More importantly, as former National Academy of Engineering President William Wulf points out, engineering teams that include people with diverse backgrounds and viewpoints produce better solutions: "Every time an engineering problem is approached with a pale, male design team, it may be difficult to find the best solution, understand the design options, or know how to evaluate the constraints" (Wulf 1998, 8).

While science and engineering are often mentioned in the same breath, engineers are not scientists. The two have very different educational and career paths and professional cultures. While most scientists are not considered to be a true member of their profession unless they have a doctoral degree, most engineers enter the workforce and pursue careers with only a bachelor's degree, although master's degrees are becoming more common in recent years. Scientists primarily work in research, investigating new areas of knowledge, whether in a university or a corporate or government laboratory. Engineers are primarily employed in private industry, in a wide variety of settings, from design to manufacturing to sales, applying scientific knowledge to solving human problems. Some engineers work directly for the public as consultants, others work for government agencies, and, of course, some are academics, conducting research and educating the next generation, but the majority of engineers in the United States are employed in corporations.

Engineers themselves are not a homogeneous group. In the nineteenth century, engineers built roads, canals, and railroads, steam engines and locomotives, and electrical power generating plants and transmission systems. In the twentieth century they built oil refineries, chemical plants, food-processing and pharmaceutical production facilities, and, as the century wore on, telephones, radios, televisions, and computers, automobiles and airplanes. In all of these fields, little noticed, women created places for themselves. Sometimes with the help of fathers, brothers, or husbands, sometimes over their objections, women engineers persevered. They were a hardy lot.

This anthology collects both profiles of pioneering women engineers—some well known in their fields, others largely forgotten—and analyses of the status of and prospects for women in engineering at various times during the twentieth century. While the first volume emphasizes individual women and their experiences as engineers, I have also attempted to include some broader perspectives on the engineering profession, its role in society, and women's roles in both. The second volume presents examples of efforts to encourage young women to pursue engineering careers and analyses of various aspects of the culture of engineering of particular relevance to women. Missing from this collection are works that specifically address the experiences of women of color, lesbians, and immigrant women in engineering. This is not a result of intentional exclusion, but of lack of material. Minorities within a minority, these women still await the serious attention they deserve.

These collections are intended to appeal to engineers themselves, both male and female, as well as others interested in engineering, women, and the role both played in twentieth-century America and continue to play today. I hope that the selections will also be of use to scholars of engineering and of professional women. My intent is not to be comprehensive, but to highlight the experiences and contributions of mostly little-known women engineers of the previous century. Women engineers continue to blaze trails, and there are many outstanding women in the profession today, but with one exception I have chosen to focus on women of the past. There are many more stories to be told about women engineers past and present and their contributions to the profession and to society. I hope this initial effort will attract more interest to the topic.

A note about format: The chapters in this collection originated in many different formats, from popular magazines to scholarly journals. The original citation systems vary widely, and have been maintained for the most part, but the text has in some cases been reformatted here for consistency and readability. Photographs of the women engineers profiled may not have appeared in the original publication. Editor's notes, where appropriate, appear italicized and in brackets [].

References

Ambrose, Susan A., Dunkle, Kristin L., Lazarus, Barbara B., Nair, Indira, and Harkus, Deborah A., eds. (1997). *Journeys of women in science and engineering: No universal constants*, Temple University Press, Philadelphia.

Commission on Professionals in Science and Technology. (2004). *Twenty years of scientific and technical employment*, STEM Workforce Data Project Report No. 1, Commission on Professionals in Science and Technology, Washington, D.C.

National Science Board. (2003). *The science and engineering workforce: Realizing America's potential*, National Science Foundation, Washington, D.C.

Rossiter, Margaret. (1982). *Women scientists in America: Struggles and strategies to 1940*, Johns Hopkins University Press, Baltimore.

Rossiter, Margaret. (1995). *Women scientists in America: Before Affirmative Action 1940–1972*, Johns Hopkins University Press, Baltimore.

Wulf, William A. (1998). "*Diversity in engineering.*" *The Bridge*, 28(4), 8–13.

Acknowledgments

Thanks to former ASCE press acquisitions editor Jack Bruggeman for suggesting the concept for this book, and to his successors Bernadette Capelle and Betsy Kulamer for seeing it through to completion. Thanks to Jill Tietjen for her encouragement and suggestions. Thanks also to the Research in Engineering Studies group at Virginia Tech for their advice and enthusiasm. Thanks to Robyn Midkiff for her valuable assistance in formatting the manuscript, and to the archivists and family members who helped with the images. Most of all thanks to my husband, Ed Champion, for his understanding and support, and to my parents, who made it all possible.

I

Historical
Perspectives

THE THREE CHAPTERS IN THIS SECTION INCLUDE A BRIEF
overview of the history of engineers and engineering, a historical look at women and engineering education in the United
States, and a social historical examination of women's participation in engineering in this country. These overviews provide context for the profiles of individual women engineers
in Section II, and background on the evolution of the culture
of the profession in the twentieth century and more recent
efforts to encourage more women to enter engineering.

In Chapter 1, Tietjen and Reynolds provide a con-
cise overview of the history of engineering and women engineers, highlighting the progress women have made in the
last 50 years. Their paper, presented at the 1999 Institute
of Electrical and Electronic Engineers International Symposium on Technology and Society, is excerpted from the
book, *Setting the Record Straight: The History and Evolution of Women's Professional Achievement in Engineering*,
and asserts that the continued low representation of women
in engineering today is due to a general lack of knowledge
about engineering as a career.

Historian Amy Sue Bix analyzes the gendered history
of engineering education in the United States, the entry of
women into engineering schools such as MIT and Georgia
Tech, and efforts to gain respect for women engineers in
Chapter 2. The author of several articles and books about
women in scientific and technical professions, Bix uses the
example of MIT to highlight the attitudes of administrators

and fellow students toward women on campus and particularly in engineering classrooms. While MIT allowed a few women to enroll each year since the late nineteenth century, they were treated as interlopers. When a women's dormitory was finally built in the early 1960s, the women were able to establish stronger support networks and work together for change. Bix describes how the imagery of corporate recruitment advertising depicts changing attitudes toward women in the profession. The article also illustrates efforts by women themselves to organize and develop networks of support in the second half of the twentieth century, while keeping an arms-length relationship with the women's movement. Chapter 2 was originally published in a special issue of the *National Women's Studies Association Journal* on "(re)gendering science fields" in 2004.

In Chapter 3, Ruth Oldenziel, a Dutch scholar of American studies, explores the intersections of class and gender in engineering, and their impact on women's participation in the profession in the United States. She discusses the professionalization and fight for status of engineering in the U.S. and its impact on women in the profession, explores the relationship between women engineers and the women's movement, compares the U.S. engineering profession with that in European countries, and describes various tactics employed by women engineers to gain legitimacy and acceptance over the course of the late nineteenth and early twentieth centuries. Oldenziel observes that many early women engineers got their start in family businesses, either following in their father's footsteps as "surrogate sons" or working in partnership with their husbands, but few associated themselves with the women's movement. She contrasts the approach of Nora Stanton Blatch, a civil engineer who was active in women's suffrage organizations of the early twentieth century, with that of Lillian Gilbreth, a pioneer in what we today call industrial engineering. Blatch, a granddaughter of Elizabeth Cady Stanton, one of the founders of the women's rights movement in the United States, sued the American Society of Civil Engineers when they refused to allow her to become a full member of the organization despite her degree and years of experience. Blatch's first husband, radio pioneer Lee DeForest, did not support her attempts to combine marriage and motherhood with an engineering career, and she eventually left both DeForest and engineering. Gilbreth, with a doctorate in psychology, partnered with her husband Frank in the development of "scientific management" and in raising 12 children. Like most women engineers of her day, Gilbreth steered clear of the women's rights movement and instead relied on hard work and stoicism to achieve recognition. Oldenziel's essay was originally published in *Crossing Boundaries, Building Bridges: Comparing the History of Women Engineers 1870s–1900s*, and received the 2002 Margaret Rossiter History of Women in Science prize for outstanding book or article on the history of women in science from the History of Science Society.

1

1999 Women Engineers Bridging the Gender Gap

Jill S. Tietjen, P.E.
Betty Reynolds, Ph.D.

The number of women receiving undergraduate engineering degrees did not reach 1% nationwide until 1972. Today, the engineering workforce is estimated to be just under 10% female and about 20% of undergraduate engineering degrees are awarded to women. This scarcity of women in the engineering field is due to a number of factors leading back to the genesis of the engineering field. This paper traces the development of engineering, the evolution of engineering into a profession, the educational requirements associated with engineers as the field evolved, early women engineers, the status of women in the engineering field today, and the outlook for women in the engineering field in the future.

Tietjen, Jill S., and Reynolds, Betty. (1999). "Women engineers bridging the gender gap." *Women and Technology: Historical, Societal, and Professional Perspectives*, Proceedings of the 1999 International Symposium on Technology and Society, IEEE, New Brunswick, N.J., 206–210. © IEEE. Reprinted with permission of IEEE and the authors.

The Genesis of Engineering

Engineering originated in the fifteenth century as a means of describing the military endeavor of designing mechanical devices for warfare. Scientific principles were applied, for example, in launching projectiles and determining approximately where they would land. Because of its birth in the military, women were automatically excluded from engineering (Ambrose et al. 1997).

After the Renaissance, engineering moved away from purely military applications, into the civilian sphere (hence the term "civil engineer") and began to resemble the practice of engineering as known in the twentieth century. Early examples of the application of engineering principles included the construction of the Italian canals, and the design and building of roads and bridges in France.

To teach engineering, formal schools were established. The Ecole Nationale des Ponts et Chaussees (National School of Bridges and Roads) was established in 1747 in France as the first formal school of engineering (Ambrose et al. 1997). In the nineteenth century, the practice and focus of engineering was significantly expanded with the development of Newtonian mechanics and the development of the steam engine (Kass-Simon and Farnes 1990). This further increased the need for formal education.

In the U.S., the U.S. Military Academy at West Point was established in 1802 to educate engineers. The oldest surviving nonmilitary engineering school in the U.S. is the Rensselaer Polytechnic Institute, established in 1824 (Ambrose et al. 1997). At the time of their establishment, and for many years after, neither school admitted women.

Professionalization of Engineering

By the end of the nineteenth century, civil and mechanical engineering were firmly established as engineering disciplines and electrical and chemical engineering as specialties followed closely behind (Kass-Simon and Farnes 1990). Many women, and men, were either officially apprenticed or learned the engineering "trade" through on-the-job training and not through formal education. However, as the engineering societies were founded (such as the American Society of Mechanical Engineers in 1880), they established requirements for membership. One of the membership requirements reflected the "professionalization" that was occurring at that time around the country and in various areas of endeavor. In order to become a member of an engineering society, a prospective member must have received a degree from an accredited school—yet most of the institutions from which an accredited degree could be acquired did not admit women. This effectively excluded women early on from membership in the engineering professional societies and obviously from a formal engineering education as well.

Engineering Colleges Admit Women

The colleges in the western and mid-western U.S. opened their doors more readily to women than the East Coast institutions. Elizabeth Bragg, the first woman to obtain an engineering degree at an American university, graduated in civil engineering from the University of California at Berkeley in 1876. In 1893, Bertha Lamme graduated from Ohio State University with a degree in mechanical engineering with an option in electricity (Kass-Simon and Farnes 1990). She is cited as the second woman to receive a degree in engineering in the U.S.

The educational situation for women in engineering improved by the turn of the century. By then, many of the western and mid-western schools and a handful of eastern universities, including MIT and Cornell, were admitting women.

Early Women Engineers

Many of the early women engineers, who are well known, were not formally educated as engineers. Ellen Henrietta Swallow Richards was the first woman elected to the American Institute of Mining and Metallurgical Engineers [For more on Richards, see Chapter 5. Ed.]. Emily Warren Roebling probably became the first field civil engineer when she took over the construction and oversight of the Brooklyn Bridge [Chapter 6]. Kate Gleason began mechanical engineering studies but was called back to work in the family gear business [Chapter 7]. Lillian Moller Gilbreth was instrumental in the founding of industrial engineering and was the first woman elected to the National Academy of Engineering [Chapter 5]. Mary Engle Pennington was actually educated as an engineer and made significant contributions to refrigeration.

[. . .]

Women and the Professional Societies

A few women in the nineteenth century were elected to membership in the engineering professional societies, primarily as "associate" members. However, it was not until 1927 that Elsie Eaves became the first woman to obtain full membership rank in the American Society of Civil Engineers [*Chapter 10*]. As late as 1942, there were only three women members of the American Institute of Electrical Engineers (AIEE) among thousands of men. Edith Clarke was the first woman elected a Fellow of the AIEE, a predecessor to the IEEE, in 1948 (Kass-Simon and Farnes 1990) [*Chapter 9*]. Lillian Gilbreth was not admitted for many years to membership in the Society of Industrial Engineers in spite of the fact that she had been instrumental in developing the field of industrial engineering. She was the first woman elected to the National Academy of Engineering; an event that did not occur until 1965.

Impacts of World War I and World War II

During World War I, women were encouraged to participate in the work force and support the war effort. Many historians believe that the vote for women's suffrage, ratification of which was finally achieved in 1920, was in gratitude for women's efforts during the war. However, in spite of that gratitude, women were not allowed to keep the jobs they had filled during the war and further demonstrate their capability to perform. The situation was so severe for women in engineering that according to the *American Men of Science* directory for 1921, there were no women at all in engineering.

Then came the stock market crash in 1929 and the ensuing Great Depression. Jobs in general were scarce and what jobs were available were unlikely to be filled by women. By 1938, the percentage of women in engineering represented less than one half of one percent of the total engineers.

World War II again presented opportunities for women to assist in the war effort personified by Rosie the Riveter. In fact, women were encouraged to enter the workforce. When the war ended, 7.8 million veterans, primarily men, came home expecting to be gainfully employed. Most of the women who had been employed found themselves no longer welcome in the workforce.

Women Engineers Since the 1950s

The launching of Sputnik in 1957, spurred the U.S. to actively encourage scientific and technical education for women and men. In spite of the encouragement to study these fields that women received, getting a job on the other side of the education still proved difficult.

Women were not enrolling in engineering colleges in great numbers. In 1972, the 525 women receiving B.S. degrees in engineering nationwide constituted only 1.2% of those graduating. In 1968, the number of women who received PhD degrees in engineering amounted to a grand total of 5 (0.2%) (Engineering Workforce Commission 1998).

During the woman's movement that began in the 1960s and continued in the 1970s, scientific and technical women pushed agendas of specific interest to them including the creating of part-time positions, better maternity leave policies, and affirmative action (Ambrose et al. 1997). By 1997, in most instances anti-nepotism rules had been eliminated.

The woman's movement has been successful in increasing the interest in women in pursuing a technical education. Women are now enrolling in engineering undergraduate curricula at levels that have never previously been achieved. In 1997, 12,160 women received B.S. degrees in engineering nationwide, constituting 18.7% of the degrees awarded. This was the highest number and the highest percentage of women undergraduate engineers ever seen. Also in 1997, 5,800 (19%) of M.S. degrees in engineering were received by women and 835 (12.2%) of all PhD degrees.

Women in Engineering vis-à-vis Women in Other Professions

There are still fewer women in engineering as opposed to any other profession. More women are pursuing medicine, law, and accounting and more women are in the scientific fields than have chosen engineering. Since 1987, the number of women graduating with undergraduate degrees in accounting has equaled or exceeded the number awarded to men. The number of women receiving law degrees climbed to 40% of the total by 1984, and by 1994, parity had been reached in first-year law school classes. About 43% of the law degrees awarded in 1995 went to women. The number of women receiving degrees in medicine was almost 42% in 1997.

Clearly, women are pursuing a wide variety of careers that in many cases require graduate education and are highly demanding. However, the number of women choosing engineering is clearly lagging the number choosing fields such as law, medicine, and accounting.

Issues Relative to Increasing the Number of Women in Engineering

We would assert that the biggest problem limiting the enrollment of women in engineering is the general public's lack of knowledge about engineering. When people opine about what engineering is and what engineers do, they sometimes think of someone who drives a train. Even more likely, they have a picture in their head of a scientist, and, of course, the scientist is male. Unfortunately this person is also somewhat "nerdy" and not necessarily well able to fit into general society. Obviously, this is not a positive role model for young women, or really for young men.

The Harris Poll conducted in 1998 reported that engineering remains a stealth profession among women and minorities. Almost 80% of women in the U.S. either are not very well informed about engineering and engineers or not at all well informed about engineering and engineers ("American Perspectives on Engineers & Engineering" 1998). This in turn implies that mothers cannot encourage their daughters to pursue engineering careers because of their own lack of knowledge about engineers and engineering.

Other significant issues were identified for the 1999 Summit on Women in Engineering hosted by the National Academy of Engineering (NAE) in May of 1999 and categorized by four transition points in a person's life: (1) elementary and middle school (encouraging membership in the math and science talent pool), (2) junior and senior years in high school (choosing engineering as a major), (3) junior year in college to three years in the work force (entering the engineering workforce), and (4) 5–7 years in the workforce (advancing in an engineering career). The issues by transition point are:

Encouraging Membership in the Math and Science Talent Pool

- Guidance on engineering career options
- Understanding of what engineers do
- Visible and effective role models
- Participation in math and science courses

Choosing Engineering as a Major

- Perceptions of engineering education and careers
- Academic policies and practices
- Motivational career information
- Visible and effective role models

Entering the Engineering Workforce

- Life and work balance
- Networks and support, including mentoring
- Career preparation in college curriculum
- Valuing diversity

Advancing in an Engineering Career

- Life and work balance
- Environmental factors including isolation, exclusion from networks, and lack of role models and mentors (National Academy of Engineering 1999)

Interestingly, themes appear as to issues at each transition point in a women's life. The most significant ones appear to be lack of role models, lack of mentors, life and work balance, and networking and support (including isolation). Although some of these issues may be common to all women in the workplace, they are exacerbated by the few women found today in the engineering workforce.

The NAE envisions significant initiatives will be undertaken through partnerships and collaborations based on the identification of the issues and the Summit itself. The most critical issue appears to be what is characterized as the "image challenge"—informing the public at large about what engineering is and what engineers do. With this knowledge, parents will be able to encourage both their sons and their daughters to consider an engineering career. The U.S. will need women engineers to be able to be competitive in a global economy in the twenty-first century and in the information age.

Conclusion

As the U.S. looks toward the future and the globalization of the economy, it will need all the talent it can find in the technical fields. Women are more prevalent today in the engineering field than they ever have been. However, many issues require resolution in order for the young women of the twenty-first century to become aware of the many opportunities available for engineers and the satisfaction that comes from an engineering career.

References

Ambrose, Susan A., Kristin L. Dunkle, Barbara B. Lazarus, Indira Nair and Deborah A. Harkus. *Journeys of Women in Science and Engineering: No Universal Constants*. Philadelphia: Temple University Press, 1997.

Kass-Simon, G., and Patricia Farnes, Editors. *Women of Science: Righting the Record*. Bloomington, Indiana: Indiana University Press, 1990.

Engineering Workforce Commission of the American Association of Engineering Societies. "For Engineering Education, 1997 Outputs Looks Like 1996," *Engineers*. Volume 4, Number 1, January 1998.

"American Perspectives on Engineers & Engineering." A "Harris Poll" Pilot Study conducted for the American Association of Engineering Societies, July 1998.

National Academy of Engineering. *The Summit on Women in Engineering*. May 17–18, 1999, Washington, D.C. Program Book.

2004   *From "Engineeresses"*
to "Girl Engineers"
to "Good Engineers"

A History of Women's U.S.
Engineering Education

Amy Sue Bix

THROUGHOUT THE FIRST HALF OF THE 20TH CENTURY AND
into the second, women studying or working in engineering were popularly perceived as oddities at best, outcasts
at worst, defying traditional gender norms. During the
last half of the 20th century, activists fought to change
that situation, to win acknowledgment of women's ability
to become good engineers. To gain public recognition for
women engineers, advocates celebrated their successes in
the field. To improve the climate for women in education
and employment, activists organized to call attention to
problems and demanded change. To aid women directly,
female engineers created systems of social, psychological,
and financial mutual support. Through such strategies,
conditions for female engineers changed noticeably over
just a few decades, although many challenges remain.

Bix, Amy Sue. (2004). "From 'engineeresses' to 'girl engineers' to 'good
engineers': A history of women's U.S. engineering education." *National
Women's Studies Association Journal,* 16(1), 27–49. © *NWSA Journal.*
Reprinted with permission of The Johns Hopkins University Press.

Engineering education in the United States has had a gendered history, one that until relatively recently prevented women from finding a place in the predominantly male technical world. For decades, Americans treated the professional study of technology as men's territory (Bix 2000b; Ogilvie 1986; Rossiter 1982, 1995). Until World War II and beyond, many leading engineering schools, including Rensselaer Polytechnic Institute, Georgia Institute of Technology, and California Institute of Technology, remained closed to women. The few women admitted to Massachusetts Institute of Technology (MIT) struggled against a hostile intellectual and social environment. Women studying or working in engineering were popularly perceived as oddities at best, outcasts at worst, defying traditional gender norms. As late as the 1960s, women still made up less than 1 percent of students studying engineering in the United States, and critics either dismissed or ridiculed women's interest in the profession. Throughout the last half of the 20th century, activists fought to change that situation, to win acknowledgment of women's ability to become good engineers.

The reasons for the strongly masculine connotations of engineering work stem, to a significant extent, from its distinctive origins. Throughout much of the 19th century in the United States, it was relatively rare for practitioners to have earned a formal engineering degree. Instead, individuals acquired credentials through on-the-job experience in a machine shop, railroad yard, or surveying crew. Such work environments excluded most women. More than that, many engineering chores involved hard, even dangerous, physical exertion, something perceived as inappropriate for respectable women. Other trends reinforced the masculinity of engineering; in the 20th century, makers of erector sets and model trains marketed these technological toys specifically as a way of turning boys into future engineers. Girls who expressed technical interests were often steered instead into the science side of home economics (Bix 2002; Oldenziel 1999; Purcell 1979, 2000; Wajcman 1991).

"Engineeresses"

In the late 1800s and early 1900s, a handful of women ventured into engineering studies, primarily at land-grant institutions (Goff 1946; Ingels 1952; LeBold 1998; Trescott 1990). For instance, Olive Dennis earned her civil engineering degree in 1920 from Cornell, then worked more than 20 years at the Baltimore and Ohio railroad. She served the B&O as a "draftsman" (as she described herself), designing some of the railroad's terminals and (more in line with stereotypes about feminine nature) designing china used in train restaurants. (Dennis 1948; Handy 1940) *[Dennis is profiled in Chapter 10. Ed.]* Women such as Dennis attracted a certain attention, since they were a rarity, a curiosity. Commenting on that female presence, a 1920s newspaper headline read, "Three Coeds Invade Engineering Courses and Compete With Men at Cornell University: Stand Well in Their Studies" (*Cornell Daily Sun* 1937, 1). The term "invade" turns up repeatedly at a number of schools in popular references to enrollment of a few female engineering students during the 1920s and again in the1940s. That word's common use following World War I and during World War II is not surprising, but does underline the extent to which women in engineering appeared as the other, entering a field that everyone assumed was and must be male territory (Bix 2000a).

The issue of women venturing into strange space came to a head with World War II, when the United States suddenly faced a "manpower" crisis. As essential men were called up for service, industry desperately needed people at drawing boards and in engineering shops to keep planes, tanks, and other war material rolling off the assembly line. There simply were not enough male engineers available. Just as manufacturers turned to "Rosie the Riveter" on the shop floor, companies sought to hire female engineers. But of course, managers immediately encountered the obvious difficulty—they could not find enough women trained for technical work.

Companies such as General Electric began recruiting women who possessed at least basic math and science skills, then gave those women emergency crash courses to turn them into wartime engineering aides. In one of the largest and most elaborate of such plans, in 1942, the Curtiss-Wright airplane company announced its initiative for training what they called "Curtiss-Wright Cadettes." Seven colleges—Cornell, Iowa State, Minnesota, Penn State, Purdue, Rensselaer Polytechnic Institute (RPI), and the University of Texas—agreed to work with the firm and teach its specially prepared curriculum to more than 600 women. Program representatives recruited sophomore, junior, and senior coeds through advertisements in college papers, calling especially for those with training in mathematics at least through algebra. Candidates underwent a 10-month immersion in classes on engineering mathematics, job terminology, aircraft drawing, engineering mechanics, airplane materials, theory of flight, and aircraft production. After that intensive exposure, Curtiss-Wright assigned Cadettes to plants to work in airplane design research, testing, and production (War Training Programs 1945).

Six of the seven campuses participating in the Curtiss-Wright program already had women enrolled. Granted, Cornell or Iowa State coeds usually majored in teaching or home economics rather than engineering, but at least students and faculty were accustomed to seeing women around campus. At these schools, announcement of the Cadette program elicited some joking about the notion of female engineering students. But Cadettes could claim to be doing their part for the war effort and on that patriotic ground, they were welcomed. By contrast, at all-male RPI, the arrival of "engineeresses" created a culture shock. Local newspapers carried giant headlines, "RPI Opens Doors to Women: Institute Breaks 116 Year Old Rule Due To War Need . . . Curtiss Wright Women . . . Invade RPI Campus" (*Rensselaer Polytechnic* 1943). Soon RPI discovered advantages to having "Katie Kaddettes" on campus. Cadettes threw themselves into school culture, joining the acting troupe and cheerleading squad. The Curtiss-Wright story represented a perfect wartime morale-booster: Cadettes proved temptingly photogenic, and *Life* published a special feature. The "engineeresses" were a curiosity, but acceptable as a temporary war measure (Bix 2000a).

"Girl Engineers"

As World War II drew to a close, returning male veterans flooded into American engineering programs, and the wartime emergency rationale for encouraging women to develop their technical talents vanished. More than that, conservative gender modes of the postwar decades brought a prevailing expectation that women's career ambitions must give way to the goal of marrying and raising children. Young girls who did express technical interests were often deliberately discouraged by negative remarks from family or teachers (Bix 2000a).

At places where engineering's macho culture had become most ingrained, such as Georgia Tech, talk of women engineers seemed ridiculous. The college humor magazine ran entire issues poking fun at the very notion, with cartoons depicting shapely women getting their curls caught in equipment, using a hydraulic testing machine to crack nuts, and invading the campus pool where men enjoyed skinny-dipping. Nevertheless, the issue of coeducation came to the fore in 1948, with rumors that Atlanta women's groups were "raising funds with which to carry through the courts the question of compelling the regents to admit women to Georgia Tech" (Van Leer 1948b). Georgia Tech President Blake Van Leer observed, "The Attorney General tells me that if they do, they are certain to win" (1948b). A test case loomed, centered around a technically talented high school woman who wished to study engineering at Georgia Tech. School president Van Leer commented:

> My personal feelings sway me in one direction, whereas my official position influences me another way. I have been associated with coedu-

cational institutions practically all of my life, and I have always felt it was wrong to discriminate against a student because she happened to be a woman. I feel that way about Miss Bonds. She is obviously a Georgia citizen and a qualified and responsible engineering student; this makes it seem wrong . . . for her to be denied an engineering education in her native state simply because she is a woman. On the other hand, Georgia Tech is traditionally a man's school. The majority of students, faculty, and alumni are opposed [to coeducation]. (1948a)

Reportedly under the influence of his wife and daughter (who both had technical interests of their own), Van Leer followed his personal inclination and started advocating women's admission. Van Leer pointed out that Georgia Tech had plenty of physical room to house women, who had already been admitted to night school and extension classes. Meanwhile, the Women's Chamber of Commerce of Atlanta passed a resolution calling on the state to let female students enter Georgia Tech. Regents immediately objected, "Here is where the women get their noses under the tent . . . We'll have home economics and dressmaking at Tech yet" (*Atlanta Journal* 1952). One later explained, "I didn't want to see . . . Tech [become] a campus full of 'debutantes' looking for a husband. I still feel very strongly that too many girls at Tech will all but destroy the seriousness of purpose in the lives of many young men at school" (Arnold 1961). In 1952, over such resistance, the board passed a measure admitting women to Georgia Tech under limited conditions (Bix 2000a).

The decision roused protests among students and alumni who felt passionately loyal to Georgia Tech's engineering-school traditions, which they regarded as inherently and necessarily male-only. The first coeds to appear on campus caused a sensation; papers published photos showing women trying on traditional freshman "rat caps." The *Atlanta Constitution* ran a cartoon showing lingerie drying on a clothesline strung from the main campus tower. The girl engineers were a curiosity; as one article explained, "A petite blonde is the first woman to attempt invasion of the home of the 'Ramblin' Wreck' since that male stronghold became coed last week. She is golden-haired Mary Joan Coffee . . . who makes it plain that she is going to Tech strictly to study and not to look for boy friends" (McNatt 1952, 1).

Such comments made it evident that female engineers of the postwar years would need to fight to be taken seriously, and in that battle, would need to band together as allies. In 1946, about 20 female engineering students at Iowa State organized a local group called the "Society of Women Engineers" to assist "in orienting new women students in the division" (*Iowa Engineer* 1946, 222). That same year, female students at Syracuse and Cornell vented their frustration at being excluded from several major engineering honor societies (or restricted to a "woman's badge" instead of full membership). The new honorary society they created, Pi Omicron, soon established chapters at colleges and universities around the nation, where members held orientations to welcome new female engineering majors. The mission was "to encourage and reward scholarship and accomplishment . . . among the women students of engineering . . .; to promote the advancement and spread of education in . . . engineering among women" (*Cornell Engineer* 1946, 14).

In 1950, female engineers in New York, Boston, Philadelphia, and Washington, D.C. began gathering on a semi-regular basis, officially incorporating in 1952 as the Society of Women Engineers (SWE), a professional, nonprofit educational service organization. The organization defined its objectives as aiming "to inform the public of the availability of qualified women for engineering positions; to foster a favorable attitude in industry toward women engineers; and to contribute to their professional advancement; to encourage young women with suitable aptitudes and interest to enter the engineering profession, and to guide them in their educational programs" (SWE 1953). One of SWE's first steps was to establish a Professional Guidance and Education Committee, which poured enormous attention into reaching potential converts. Members personally wrote to dozens of high school girls, sending information about engineering and replying to questions. In 1958, Boston's chapter of SWE published a pamphlet containing biographical sketches of a few "typical" women engineers and explanations of how girls could prepare to enter engineering. SWE's authors concluded, "If this pamphlet shall have inspired one young woman to consider an engineering career . . . and one parent to 'encourage' the daughter's desire to enter the technical field, this pamphlet will then have been a worthwhile venture" (Miller 1964).

Such volunteer guidance reflected one of SWE's primary beliefs: that girls often shied away from technical pursuits because they simply did not realize that women could and did go into engineering. Irene Carswell Peden, associate professor of electrical engineering at the University of Washington, wrote:

> It is important to think of women engineers as real people doing real jobs which the student could do, too. . . . A girl is not likely to choose a career field disapproved by her parents, teachers, classmates, and friends. All of these people . . . seem to be responding in part to an erroneous but popular image of the woman engineer as a cold, . . . aggressive female who trudges through life in her flat-heeled shoes without a

man in sight (away from the job). . . . Many women engineers are very attractive; most represent a perfectly normal cross section of femininity. The only way that this image can be brought into line with reality . . . is by . . . personal contact. (1965, 2)

In 1954 and 1955, members of Cleveland's SWE appeared on local television programs as living proof of women's small but persistent presence in the engineering profession. At a time when many Americans perceived female engineers as odd manlike creatures, SWE representatives took pains to offer a presentable feminine image, emphasizing that many of them were married and had children. Advocates believed that women engineers could gain greater acceptance in society simply by making themselves more visible in a professional, positive way (Bix 2000b).

In the mid-1950s, SWE expanded its efforts to include more active outreach. Members volunteered to assist at "Junior Engineer and Scientist Summer Institute" (JESSI) programs which brought high schoolers to college to explore science and receive educational guidance. At one JESSI session in Colorado, 53 girls listened to a five-woman panel discuss why they had chosen engineering careers. Female engineers led JESSI students on visits to industry and gave the girls (and boys) tours of their laboratories (Rutherford 1954).

By 1957, female engineering students at Drexel, Purdue, University of Colorado, City College of New York, Missouri, and Boston had founded student sections of SWE, and the parent organization welcomed its junior counterparts. Established women engineers vividly remembered how intimidating it felt to be the sole woman in an engineering class. They knew, as Helen O'Bannon wrote, that "being one of a small group following a path that appears to violate society's norms is lonely" (1975). Mildred Dresselhaus argued that such young women deserved support from older mentors, who could provide the encouragement necessary to "keep going when the going gets rough or when [a girl] begins to ask, 'Is it worth it?'" Successful role models could give new students a boost in confidence, a chance "to see by example that women can 'make it' in engineering" (1975a). Older professionals especially sympathized with those young women just entering Georgia Tech, and in 1958, Atlanta's SWE chapter sent several members to participate in Georgia Tech's start-of-the-year camp for first-year women. "One must realize that there are this year approximately 1,300 freshmen at Georgia Tech and only 19 freshman coeds. There will be numerous problems and SWE Atlanta Section is proud to play an integral part in the quite difficult assimilation of female engineering students in an almost all-male school" (Dresselhaus 1975b, 30).

"Good Engineers"

More than simply pressing all-male schools to admit coeds, more than just encouraging young women to consider engineering studies, SWE and other advocates wanted to convince doubters that women could be good engineers. They strove to win respect, knowing that opposition remained quite visible. In 1955, Eric Walker, Penn State University's engineering dean, wrote a column saying, "Women are NOT For Engineering." Walker declared that most women did not have the "basic capabilities" needed for engineering. He concluded that teaching them didn't make sense; since "[t]he most evident ambition of many women is to get married and raise a family . . . few companies are willing to risk $10,000 on a beautiful blonde engineer, no matter how good she may be at math" (1955, 18).

The effort needed to counter such critics can be seen in the story of women at the Massachusetts Institute of Technology. The school had actually been coed since 1871, and between the 1920s and 1940s, MIT averaged 50 female students on campus each year, amidst approximately 5,000 men. Coeds represented a curiosity. The student newspaper introduced a 1940 class member as a New York "glamour girl" who hoped to work in cancer research and won a $100 bet from fellow debutantes by gaining admission to MIT. Officially, coeds remained invisible. President Karl Compton told incoming students, "In choosing MIT, you've taken on a man-size job" (MIT 1941, 5). Campus traditions represented masculinity itself; as an official welcome, the institution held a "smoker" for freshmen and their fathers. Initiation took place at MIT camp, featuring water fights with the sophomores, baseball games with faculty, and plenty of male-bonding rituals. Engineering programs seemingly presented problems for inclusion of coeds. Civil engineering students learned surveying and other field techniques at a rough camp, with accommodations judged unsuitable for females. Mechanical engineering class required round-the-clock observations of engine performance; generations of male students turned the "twenty-four-hour boiler tests" into beer parties. The prospect of women staying overnight with men in the lab seemed inappropriate (Bix 2000b).

In the years following World War II, MIT kept admitting a few women per year, then proceeded to ignore their existence as a minor anomaly. In 1947, the dean of students defined MIT as intended "to prepare men for . . . engineering, . . . educate . . . men for responsible citizenship" (Baker 1947). Throughout the foreseeable future, coeds would "continue to be grossly outnumbered by men in classroom and lab," officials admitted. As Florence Stiles, adviser to women students, explained, the sense was that "women in general do not make acceptable engineers" (1946). One observer later wrote, "Before 1960, women entered MIT at their own risk. If they succeeded, fine; if they failed—well, no one had expected them to succeed" (Wick 1970). The few coeds around hesitated to rock the boat. One of MIT's female graduates from this era remembered:

> I was very conscious of having to represent women in each class. If I did anything wrong, . . . said anything stupid, it would be ammunition for all the men who didn't want us there in the first place. . . . Discriminatory events were so common that it didn't occur to us to object"; [besides], "other engineering schools weren't accepting women . . ., so even though MIT was only accepting twenty a year. . . . I felt MIT was doing us an enormous favor to have us there at all. (Jansen 1977)

Through the mid-1950s, many at MIT argued for ending coeducation, citing the high female drop-out rate. Margaret Alvort, women's-house supervisor, wrote that her "doubt as to whether [coeds] belong . . . has grown into certainty that they do not." If MIT wanted to serve the nation by graduating as many top-notch engineers and scientists as possible, then "there is little in the records of the girls . . . to justify their continuance" (Alvort 1956). The school medical director agreed, "[W]hen there is such a shortage of engineers, one wonders if we are justified in taking positions away from male students for female." Coeds might bring "pleasure and ornamentation" to campus, but usually proved unable to hold their own against MIT men's competitiveness and "high-grade intellects." In short, he concluded, "except for the rare individual woman, [MIT] is an unsuitable place" (Keller 1981, 12).

Significantly, MIT president James Killian believed some women could succeed in scientific and technical fields and therefore argued for their continued access. He wrote, "I do not see how the Institute, having admitted women for so long, can now change," nor should it, considering that America's Cold War race with the Soviet Union called for development of all professional talent. Striving to "think more boldly . . . about recognizing [women's] presence," Killian sought support for coeducation (1956). In 1960, alumna Katharine Dexter McCormick pledged $1.5 million to build MIT's first on-campus women's dorm. McCormick knew that in her day, MIT had enrolled 44 women, a figure that had barely risen five decades later (Bix 2000b).

Dedication of McCormick Hall in 1963 attracted national publicity. "Hardly anyone imagines girls attending mighty MIT," *Time* reported. "Yet last week Tech . . . dedicated its first women's dormitory to go with its first women's dean, an attractive blonde lured from nearby Radcliffe." MIT used McCormick Hall's opening to draw attention to its female students. Noting that "opportunities for women in science [and] engineering . . . are clearly increasing," 1963's catalog mentioned up front that MIT was coed. Women's applications jumped 50 percent in 1964. Backers of coeducation hailed McCormick Hall as a "vote of confidence, testimony . . . that women are to remain a permanent part of MIT" (1963, 51).

Now that the university had finally created a physical place on campus for female students, women's dean Jacquelyn Mattfeld called on MIT to integrate coeds intellectually and socially. A "conservative . . . Wall Street attitude toward women still runs through MIT's veins," she declared. Many male professors and students regarded female undergrads as "incompetent, unnatural, and intruders" (1965).

Rather than waiting and hoping for such hostility to vanish, a new generation of MIT coeds began to band together to consider remedies. They began addressing issues such as employment discrimination, after facing corporate interviewers who openly questioned how long a woman engineer would remain on the job, doubted whether she could think about mechanical details "like a man," and offered distinctly lower salaries. To discuss such problems, the school's newly invigorated Association of Women Students (AWS) helped organize a "Symposium on American Women in Science and Engineering" at MIT in 1964. Planners hoped to attract widespread media coverage, teaching industry, the public, and young women themselves that women could be good engineers and scientists. The symposium attracted college faculty and administrators, high school students and guidance counselors, plus more than 250 delegates from Smith, Radcliffe, Wellesley, University of California, Georgia Tech, Northwestern, Purdue, and other institutions. The then-novel coming-together of such a large group served an important purpose in itself; one mechanical engineer from Michigan State University found it "reassuring to see so many other women in the same situation" (*Michigan State News* 1964, 1). Speakers such as University of Chicago professor Alice Rossi called on society to encourage independence, curiosity, and reasoning power in girls, while Radcliffe president Mary Bunting called on employers to provide day-care and flexible schedules to help women balance motherhood and work (Bix 2000b).

Corporations of that period did not prove particularly proactive in initiating accommodations for career women. Indeed, through most of the 1960s, company recruitment for engineers ignored the existence of women in the profession. A 1954 advertisement in MIT's engineering magazine for the Ramo-Wooldridge communications corporation featured a photograph of its senior staff, 23 "key men" clad almost identically in dark suits and white shirts. A 1968 ad centered around a photograph of a serious-looking middle-aged man in white shirt, dark tie, and thick glasses, under the headline, "This is the image of a Kodak mechanical engineer." Such advertisements reinforced the attribution of mechanical interests and aptitude to men, while relegating women to the sidelines. In one 1958 ad for the Avionics Division of ITT, a cartoon showed a little boy building a complicated machine, while a girl carrying a doll looked up admiringly and asked him, "Have you always been a genius?"

In attempting to catch the attention of engineering majors, companies pandered to the masculine identity and image of their profession. The University of Michigan's engineering magazine ran a 1966 advertisement for Douglas Aircraft that showed a drawing of a young man floating happily along a beach in an inner tube, while two young women in skimpy swimsuits gazed at him adoringly. Such appeals often crossed the line into sheer sexism. Another Douglas ad, under the headline "Intrigued by Exotic Designs?" featured a cartoon of a man sitting at a drawing board, carefully laying out an illustration of a naked woman outstretched like an airplane in flight, with measurements accenting the curves of her breasts, legs, and behind. One advertisement for Chance Vought Aircraft used a bizarre illustration of a young man having an intimate meal with a female robot, complete with flirtatious eyes and pointed breasts. The ad copy read, "Do mechanical brains intrigue you? Do those intellectual vamps arouse your engineering instincts? Then why go on ogling? Especially if you're an electrical or mechanical major! Plan to enjoy the company of the best mechanical computers. Create your own electronic brains for missile guidance."

Within such a recruiting climate, the equation of "engineer" with "male" amounted to a real obstacle for female professionals. Despite the fact that the 1960s were characterized by a national shortage of qualified engineers, employers often refused even to consider women candidates. According to *Industrial Relations News* in 1961, interviews with personnel administrators and corporate managers revealed that "81 percent wouldn't hire female engineers, and most of the remaining respondents would be dead set against permitting them to reach middle-management levels" (*SWE Newsletter* 1961, 1).

The fact that employers expected to hire male engineers did not escape female job candidates. Gerda Kohlheb, a chemical engineering student at City University of New York in the mid-1960s commented, "Most women who do go through an engineering curriculum find a great deal of opposition during their college career. . . . [M]ale students tend to regard her as somewhat of an intruder. Upon entering a classroom, the looks that sometimes greet her are not unlike the ones she would get if she walked into a men's room" (1967, 13). In a tactic to avoid such situations, Kohlheb went abroad to seek a summer job in the chemical industry, saying, "Most American companies were hesitant to accept a woman to work in the field, and I hoped I would be accepted as a woman more readily in Europe." Kohlheb, "as extra insurance . . . decided to commit the sin of omission"; on her job application, she used only a first initial, rather than her first name (1967, 13).

Given the persistence of employers' doubts about women engineers, many activists of the late 1960s and early 1970s designed plans to help female students break through the barriers. Working in MIT's engineering school, professors Mildred Dresselhaus and Sheila Widnall inaugurated a freshman seminar aimed at acclimatizing coeds to engineering. To make women comfortable with manual skills that boys traditionally picked up from hobbies or from fathers, the syllabus included lab projects in electronics, welding, and model-building. Dresselhaus further helped organize meetings entitled "Let's Talk about Your Career," where female students consulted faculty, staff, and guests for advice on graduate school, employment, and the perpetual issue of blending marriage with work. Arguing that male students' familiarity with the business world gave them a competitive advantage, MIT alumnae started an annual seminar, "Getting the Job You Want in Industry: A Woman's Guerrilla Guide to the Pin-Striped World." By advising coeds on resume writing and interview techniques, alumnae hoped to level the playing field (Bix 2000b).

Similar efforts at promoting women's advancement took place on campuses such as Purdue University, where an intensive recruiting and retention campaign had raised enrollment of female engineering students from 46 in 1968, to 280 in 1974, to more than 1,000 in 1979, the largest such class in the nation. Purdue also had one of the country's most active student SWE chapters, which published its own newsletter, and ran a "big sister" program pairing entering women with upper-class mentors. SWE offered help in locating summer jobs and produced an annual "resume book," showcasing members' credentials, which it sold to potential employers (Bix 2000a).

Meanwhile, important changes were underway. The 1964 federal Civil Rights Act had included language barring employment discrimination on the basis of sex and created the Equal Employment Opportunity Commission. When enforcement lagged, the National Organization for Women mobilized to pressure the agency to secure women's workplace rights. Those legal and political developments had a direct impact on the official culture of engineering, transforming the tone of recruiting. Where formerly corporations had casually deployed sexist imagery and equated the concepts of "engineer" and "male," leading companies of the 1970s carefully stipulated that they were "an equal opportunity employer" (Woloch 1999).

In many such advertisements, corporations highlighted the stories of women engineers already in their employ as a means of attracting the attention of female students entering the job market. This tactic also fostered favorable public relations for the firm, bolstering an impression of compliance with equal-opportunity rules. Kodak, which just a few years before had defined the stereotypical nerd as its engineering image, ran a 1973 advertisement featuring a group photo of 15 women engineers, each identified by degrees earned, specialty, and responsibilities. A University of Minnesota mechanical engineer "designs and troubleshoots hydraulic systems, bearings, and shaft seals. She is a specialist on friction, wear, and lubrication." An electrical engineer from the South Dakota School of Mines handled "machines [that] are three stories high, a football field long, and work to the tolerances of an expensive watch in depositing emulsion layers on color film." Tongue in cheek, the ad told readers, "This picture could be misleading. Engineering jobs at Kodak are not restricted to ladies."

One of the most significant questions, of course, was whether female engineers could enter training for management and receive promotion opportunities alongside their male colleagues. General Electric promised a readiness to bring women into the higher ranks of business through programs of rotating assignments and field experience. A 1974 ad showed a drawing of a young woman smiling broadly and leaning back confidently in her executive chair, while the headline read, "We're Looking for Engineers Who Were Born to Lead."

In playing up to women's aspirations, many companies deliberately integrated feminist imagery and language into their recruitment of female engineers. One power systems corporation offered the slogan in 1975, "Women Engineers: You've Come A Long Way, But You'll Be Surprised How Much Farther You Can Go With Gibbs & Hill." Another 1975 ad, for semiconductor manufacturer INMOS, used a drawing of young boy in a baseball uniform, saying, "When I grow up, I want to be an engineer, like my Mom. . . . A new wish for a different time." Not surprisingly, such feminist sentiments were especially prominent in the advertising companies increasingly placed in the *SWE Newsletter* and SWE conference programs of the 1970s, appealing directly to the growing constituency of female professionals. Under the headline, "We make products other than appliances— and hire people other than men," GE specifically invited female "engineering and manufacturing professionals . . . to work with us in creating, manufacturing, and marketing . . . advanced, high technology jet aircraft engines. . . . You'll have the same opportunities for professional advancement as your male counterparts, the same pay and the same status" (*SWE Newsletter* 1978, 5).

Some young female engineering students just completing their studies in the early 1970s were impressed by this dramatic rhetorical shift, ready to believe that a cultural revolution in attitudes toward women engineers was underway. A sophomore majoring in chemical engineering at Penn State University declared in 1971 that "probably with the Women's Lib movement, previous discrimination will be lessened or non-existent by the time I apply for an engineering job" (*Centre Daily Times* 1971, 7). But other engineering graduates and older women were more skeptical about the likelihood of fundamental change in the workplace climate. Moreover, SWE cautioned that with a perception that companies were targeting female job candidates, "unfortunately, we're seeing a little backlash (. . . on the part of male students) which we don't like very much" (*SWE Newsletter* 1979, 1). As SWE observers were well aware, federal legislation and new recruiting language would not instantaneously convert engineering into a feminist paradise. Underneath equal-opportunity rhetoric, many female engineers ended up underemployed, working in positions that failed to use their full capacity, and frustrated by deeply ingrained discrimination (Durkin 1975; Gitschier 1973; Gluch 1977; Mathis 1972). Women engineers frequently complained that male colleagues and supervisors initially tended to doubt their ability. Thus, a female engineer was forced to defend her right to employment and demonstrate her skill through hard work, while "her male counterpart . . . is basically accepted as able to do the work successfully, unless he himself proves otherwise" (Bugliarello 1971, 8).

Activists at MIT warned that the environment on campus remained unfriendly toward women. An ad hoc committee, co-chaired by Dresselhaus and engineering major Paula Stone, drew on fundamental feminist principles to declare that "a discriminatory attitude against women is so institutionalized in American universities as to be out of the awareness of many of those contributing to it" (MIT Ad Hoc Committee 1972, 3). The report noted that women at MIT faced both open opposition and silent prejudice, concluding:

> If many people (professors, staff, male students) . . . persist in feeling that women jeopardize the quality of MIT's education, that women do not belong in traditionally male engineering and management fields, that women cannot be expected to make serious commitments to scientific pursuits, that women lack academic motivation, that women can only serve as distractions in a classroom, . . . then MIT will never . . . be a coed institution with equal opportunities for all. (MIT Ad Hoc Committee 1972, 3)

The document represented a self-directed rallying cry, telling MIT women that gender discrimination would change only when female students, faculty, and staff organized to demand improvement. The early 1970s brought a burst of activism, as MIT women drew strength from the national feminist movement to assert their presence physically, intellectually, socially, and politically. Advocates carefully listed all the awards coeds received in order to document that women could indeed be good engineers, and to show remaining doubters that women could lead and succeed in the most difficult technical studies. MIT's Admissions Office had revised photographs and text in the catalog to highlight coeds and sent special recruiting material to all female National Merit and National Achievement semifinalists. AWS feared, however, that it would require more "high-powered" efforts to increase female enrollment, to overcome social forces pushing girls away from science and engineering and "to demythify incorrect assumptions about women at MIT" (MIT Association of Women Students 1977). AWS produced its own pamphlets encouraging high school girls to apply and urged members to contact hometown seniors over Thanksgiving and Christmas vacation. "The women in particular may just need an encouraging word from you before taking the plunge" (MIT Association of Women Students 1977). MIT coeds also volunteered to sit in the Admission Office during peak interview period, ready to chat with interested girls (Bix 2000b).

To help MIT women maintain a positive sense of identity within a male-dominated atmosphere, campus women's groups initiated monthly colloquia addressing wide-ranging feminist subjects such as the nature of androgyny, sexism in popular culture, and the strengths and difficulties of two-career marriages. Dresselhaus and Wick created a new organization, the Women's Forum, which brought together undergraduates, graduate students, faculty, staff, and wives to develop "consciousness-raising skits," to express concerns about women's health, athletic opportunities, day care, and career planning. Women's dean Jacquelyn Mattfeld and her successor, Professor Emily Wick, served as administrative advocates for MIT's coeds, ready "to assist women students as they make their way through this very male institution" (Wick 1971). Mattfeld and Wick stepped in to mediate when coeds encountered trouble dealing with advisors, professors, or teaching assistants. Similarly, many of MIT's few women faculty considered it their responsibility, as successful professionals, to lobby on behalf of other women on campus. Widnall complained, "Engineers may have a view of engineering which is twenty years out of date, and they communicate that to other people. Engineers have an image of engineering that is very masculine . . . [and] takes a long time to change" (1976, 12). She described women's activism as a "very exciting" force which could open opportunities for new generations of girls. "There's obviously a direct connection between militant feminism in the junior highs and the ultimate enrollment of women in engineering. . . . Everybody, mothers in particular . . . are much more aware of the importance of encouraging their daughters to take life seriously" (13).

In 1973, MIT convened another workshop on women in science and engineering. Embracing feminist language, President Jerome Wiesner spoke about a need "to encourage women's participation in every aspect of our technological society. This is another front in the almost universal battle for equality of opportunity" (*Women in Science and Technology* 1973, 3). Commemorating the 100th anniversary of MIT's female graduates, advocates considered 1973 an occasion for celebration. Though they still saw much room for improvement, female enrollment had more than tripled over just two decades. By the late 1970s, women made up 17 percent of MIT undergrads and 12 percent of engineering majors. The sheer increase in population mattered; as women became more of a presence on campus, activists gained a critical mass for organization and for visibility. Female graduate students formed their own society, as did women at Lincoln Lab. Such groups kept women's issues on the front burner, providing an identity and a cause for many (especially valuable to female faculty and graduate students based in departments with few other women) (Bix 2000b).

MIT activists engaged the national women's movement, thinking about how their immediate interest in supporting women's engineering and science education connected with broader feminist issues. In 1974, the Women's Forum invited Gloria Steinem to speak, and her speech to a packed auditorium defined feminism as political, social, and educational autonomy for women (Brandeau 1975, 1). Activists found food for thought in Steinem's message, given that female students still frequently felt like second-class citizens. In subsequent years, MIT women would gain courage to start complaining about sexual harassment, about obscene mail sent over computers, about male colleagues who refused to take women seriously, made them feel invisible, and undermined their self-confidence. The women assumed a right to demand change.

On a broader scale, this same force of energized activism found expression nationwide by the early 1970s, as female engineers followed specific strategies to promote women's place in the field. Established professionals offered support for juniors; for instance, the Los Angeles section of SWE provided speakers and counselors to student sections at USC, UCLA, Loyola Marymount, Harvey Mudd, Cal State Long Beach, Pomona, Fullerton, and Cal Poly San Luis Obispo. Such campus SWE groups provided vital intellectual, social, and psychological support for incoming female engineering majors. Karen Lafferty Instedt, an Ohio State student from 1968 to 1971, later wrote that SWE gave her "an opportunity to meet the other female engineers who, like me, were isolated in their respective fields and classrooms. The SWE section functioned as a refuge of sorts—where one could find an understanding ear from a peer or a kindhearted, encouraging professor or dean" (1978, 1). By the end of the 1970s, student sections had been chartered in more than 170 colleges, universities, and technical institutes. SWE held an annual national student conference featuring technical sessions and exhibits, professional workshops, industrial tours, even sessions on career planning, power dynamics, management, personal assertiveness training, and how to "dress for success." By decade's end, SWE's membership totaled more than 10,000 women and men. As SWE grew, its leaders were able to mobilize outside support, collecting money to help finance college education for young women pursuing technical studies. SWE administered annual scholarship competitions for female engineering majors, funded by RCA, Westinghouse, other major companies, and women engineers themselves. By publicizing such awards and showcasing the winners, SWE sent a message that women had won a permanent place in American engineering, fully deserving of social and financial support (Bix 2000a).

Activists of the 1970s organized dozens of conferences, open houses, and other public events in many states to celebrate and advance women's achievements in engineering. Some meetings were organized to bring female students together with each other and with older mentors. For instance, the University of Washington (with almost 450 women engineering students in 1977) hosted an annual conference where those coeds met with working professionals such as Rockwell ceramics engineer Bonnie Dunbar. The SWE section at the University of North Dakota brought in corporate representatives to talk about how to project a professional image, how to have a successful interview, and how to balance work and marriage (Bix 2000b).

Other conferences were organized by college engineering women themselves for younger girls as a way of encouraging them to pursue technical interests. SWE's philosophy generally assumed that girls and boys essentially possessed similar ability to excel in math and science. They blamed girls' relative lack of interest in engineering on socialization that handed dolls to girls and toy tools to boys, that put girls in home economics and boys in shop class. SWE further attributed girls' underrepresentation in engineering to failures of the school system, finding fault with guidance counselors who failed to take girls' ambitions seriously or let them drop math and science. To counter such problems, a 1973 University of Illinois conference, titled "Women in Engineering: It's Your Turn Now," gave high school junior and senior girls a chance to participate in "rap sessions," informal conversations with college SWE members. A 1974 symposium sponsored by SWE sections at the Universities of Florida and South Florida featured a tour of the Kennedy Space Center, plus discussions of student financial aid, co-op programs, and career openings. Promotional material read:

> As an engineering student you'll gain something most women don't get in college, a professional skill which can be used immediately upon graduation . . . [with] the highest starting salary bracket of the major professional job categories for women holding a bachelor's degree. . . . You owe it to yourself to look into the possibilities and opportunities offered by engineering. (*SWE Newsletter* 1979, 4)

Other SWE chapters went directly into the high schools as self-described "missionaries," seeking to spread the message that women could be good engineers. Starting in 1976, Berkeley's SWE section sent teams to visit local junior high classes. Presenters described how they became interested in engineering and sought "to dispel myths about women in engineering" (SWE Berkeley 1980). Other SWE activists sought to influence even younger girls, to encourage curiosity and technical enthusiasm in elementary-age children. Boston's SWE published a coloring book titled Terry's Trip, the story of a girl visiting her aunt, a mechanical engineer. After talking to engineers, male and female, who worked at her aunt's toy factory, Terry concluded, "Maybe some day I'll be an engineer like Aunt Jennifer." In a similar project, North Carolina SWE produced a booklet titled Betsy and Robbie, the story of a girl who visited a university engineering fair and became fascinated with the robot designed by a female student. Such material emphasized that women were fully qualified for engineering, a discipline requiring creativity and logic more than physical strength. Photos documented the daily activities of women engineers working in industry, government, and academia, providing role models to win young women's interest and create respect for female engineers as good engineers (Sloan 1979).

Conclusion

Amidst this climate of activism, it is worth emphasizing that while many female engineers such as Widnall embraced the philosophy of feminism, others actively rejected the label (Mack 2001). Engineering tends to be a conservative field, and many women shied away from anything that might be judged as too radical. They worried about popular perceptions of "women's lib" and feared that being active in SWE would get them branded as "troublemakers." Nevertheless, such women benefited from the efforts of activists who did identify with the feminist movement and undertook conscious, passionate campaigns to break down institutional barriers.

At the end of the 20th century, however, women were still nowhere close to approaching proportional representation in the profession. In 1979, women made up 12.1 percent of undergraduates enrolled in engineering across the United States; by 1998, that percentage had gradually risen to 19.7 percent. In terms of graduation rates, in 1996, 11,316 women earned bachelor's degrees in engineering, 17.9 percent of the nationwide total. By occupation, women constituted 9 percent of all engineers in 1998 (NSF 2000).

Evidence confirms suspicions that while equal opportunity sentiment and federal legislation might have helped open some doors for some women in the 1970s, the changing language and imagery of employment ads to denote hiring diversity did not solve more fundamental problems. In 1993, SWE reported that "in some respects, women in engineering begin careers at parity with men or better, but as one looks at more experienced people, this picture changes" (*SWE Report on Employment*). A survey of more than 1,700 professionals showed that although the average annual base salary of female engineers under age 30 was actually higher than men's by a couple of thousand dollars, average male salaries surpassed women's in the 30 to 39 age group and kept rising at a faster pace. The gender pay differential among senior professionals favored male engineers by almost $15,000. SWE analysts concluded, "Women engineers appear to fare poorly. . . . [C]omparison of salaries suggests that nothing is being done about long-recognized inequities" (*SWE Report on Employment*). In a finding clearly related to pay, SWE research documented that at every age level, a greater percentage of male than female engineers had moved into management (SWE 1993). Data from the National Science Foundation (NSF) suggested that for female engineers less than 10 years after graduation, median annual salaries were $2,000 to $3,000 below those of men in the same cohort. In older generations, the salary gap widened to $5,000 or more; for example, among engineers 20 to 24 years out of school, men earned a median salary of $68,600 versus women's $60,000 (NSF 2000).

Despite the optimistic predictions of female undergraduates in the 1970s that gender bias in the engineering workplace would soon become a relic, the issue remained very much alive going into the 21st century. The SWE survey reported a distinct gender gap in perceptions of job discrimination. More than half of all male engineers surveyed, 55 percent, said that men and women were always treated equally in engineering, while less than one-third of women, 26 percent, agreed. Fifteen percent of female engineers reported that they saw consistent inequities, a statement supported by just 5 percent of men. Another 2 percent of men surveyed (and zero women) indicated that they saw "reverse discrimination" in current conditions (SWE 1993).

Ironically, it would be in engineering education itself, where advocates had pushed so hard for undergraduate women to gain access, that employment disparities remained particularly stubborn. Observers had long noted the relative absence of women in engineering departments, especially at levels above assistant professor. The National Science Foundation 2000 report confirmed such perceptions and concluded that although part of the gender gap in rank and tenure might be attributable to the relative scarcity of women earning engineering doctorates, that factor alone could not account for the difference. On occasions when a woman was promoted to dean or even department chair in engineering, the event still appeared sufficiently unusual in 2003 to make headline news (Jerousek 2003).

The reasons for female underrepresentation in academic engineering posts are numerous and complex, yet one persistent factor remains the discipline's chilly climate. In 1994, three tenured women at MIT began comparing notes on disturbing experiences. Though the professors worried about "putting a life-time of hard work and good behavior at risk . . . [and] feared being seen as radical trouble makers" (Goldberg 1999, A1) they proceeded to collect data (an approach suiting MIT's scientific mindset). As MIT's official self-examination ultimately acknowledged, female faculty had received lower financial resources, less work space, and fewer rewards than male colleagues. Engineering professor Sallie Chisholm described the phenomenon as "microdiscrimination," small but cumulative assaults on women's careers, "unrecognized assumptions and attitudes that work systematically against women faculty" (Koerner 1999, 56). The report declared:

> Marginalization increases as women progress through their careers at MIT . . . [and] this pattern repeats itself in successive generations. . . . Each generation of young women, including those who are currently senior faculty, began by believing that gender discrimination was 'solved' in the previous generation and would not touch them. Gradually, however, their eyes were opened to the realization that the playing field is not level after all. (MIT report 1999)

In working to document systematic inequities and demand change, the MIT women of the 1990s were following in the footsteps of others who had organized to improve conditions for women trying to enter engineering. In the 21st century, it will take ongoing attention to neutralize the more subtle and hence more stubborn problems of "microdiscrimination." In that fashion, the debate goes on concerning women's place in engineering, one of the most traditionally male professions.

References

Alvort 1956. Margaret Alvort to L. F. Hamilton, June 21, 1956. AC220, box 2, folder 2; MIT Archives.

Arnold 1961. Robert D. Arnold to Fuller E. Callaway, Junior; June 27, 1961; box 9, folder "coed", 85-11-01, Georgia Tech Archives.

Atlanta Journal. 1952. "Regents Vote Tech as Coed," *Atlanta Journal,* April 9, 1952: 1.

Baker 1947. Memo from Everett Baker; January 26, 1947; AC 4, box 26, folder 12; MIT Archives.

Betsy and Robbie. n. d. ca. 1983. Box #119, file "Betsy and Robbie." SWE collection, Wayne State University Archives.

Bix, Amy Sue. 2002. "Equipped For Life: Gendered Technical Training and Consumerism In Home-Economics, 1920–1980." *Technology and Culture* 43(4):728–54.

——— 2000a. " 'Engineeresses' Invade Campus: Four Decades of Debate over Technical Coeducation." *IEEE Technology and Society Magazine* 19(1):20–6.

——— 2000b. "Feminism Where Men Predominate: The History of Women's Science and Engineering Education at MIT." *Women's Studies Quarterly* XXVIII (1 & 2):24–45.

Brandeau, Margaret. "Steinem: Castes Trap Women," *The Tech,* January 15, 1975: 1.

Bugliarello, George. 1971. "Women, Technology, and Society." In *Women in Engineering: Bridging the Gap Between Society and Technology,* edited by George Bugliarello, Vivian Cardwell, Olive Salembier, and Winifred White. Chicago: University of Illinois at Chicago Circle: 1–18.

Centre Daily Times. 1971. "More Coeds Preparing to Become Engineers," *Centre Daily Times,* July 28, 1971: 7.

Cornell Daily Sun. 1937. "Three Coeds Invade Engineering Courses and Compete With Men at Cornell University." *Cornell Daily Sun,* November 12: 1.

Cornell Engineer. 1946. "Pi Omicron," *Cornell Engineer,* April 1946: 14.

Dennis, Olive, 1948. "Modernization of Railroad Passenger Facilities," *The Cornell Engineer* 14(2): 7–9, 34, 36.

Dresselhaus, Mildred S. 1975a. "A Constructive Approach to the Education of Women Engineers." Box 128, file "Women in Engineering—Beyond Recruitment Conference Proceedings, June 22–25, 1975." SWE collection, Wayne State University Archives.

Dresselhaus, Mildred S. 1975b. "Some Personal Views on Engineering Education for Women." *IEEE Transactions on Education* 18(1):30–4.

Durkin, Glen C. 1975. "Engineering—A 'Weird' Career for Women?" *Penn State Engineer,* January, 19.

Gitschier, Jane. 1973. "Sex: Female, Major: Engineering." *Penn State Engineer,* November, 31.

Gluch, Bonnie. 1977. "Women in Engineering: A Personal Perspective." *Penn State Engineer,* October, 26–7.

Goff, Alice. 1946. *Women Can Be Engineers.* Youngstown, Ohio.

Goldberg, Carey. 1999. "MIT Acknowledges Bias Against Female Professors." *New York Times*, 23 March, A1, A16.

Handy, Adelaide. 1940. "Woman Designer of Bridges Has Enhanced Rail Travel," *New York Times*, 22 December, 40.

Ingels, Margaret. 1952. "Petticoats and Slide Rules," *Midwest Engineer*, August, 2–16.

Instedt, Karen Laferty. 1978. "How Should SWE Serve Undergraduates?" *SWE Newsletter*, June/July, 1.

Iowa Engineer. 1946. "New Society Organizes," *The Iowa Engineer*, May, 1946: 222.

Jansen 1977. Interview of Christina Jansen by Shirlee Shirkow, 1977: MC86, box 9; MIT Archives.

Jerousek, Madelaine. 2003. "Engineering College at ISU Gets First Female Department Head." *Des Moines Register*, 30 May, B1

Keller, Evelyn Fox. 1981. "New Faces in Science and Technology: A Study of Women Students at MIT." August, MIT Archives.

Killian 1956. J. R. Killian, Jr. to J. A. Stratton, October 22, 1956; MIT Archives.

Koerner, Brendan I. 1999. "The Boys' Club Persists," *U.S. News & World Report* (5 April):56.

Kohlheb, Gerda. 1967. "Some Thoughts from a 'Lady Engineer'." *Chemical Engineering*, 11 September, 13–5.

LeBold, William K. and Dona J. LeBold. 1998. "Women Engineers: A Historical Perspective," *ASEE Prism*, March:30–2.

Mack, Pamela. 2001. "What Difference Has Feminism Made to Engineering in the Twentieth Century?" In *Science, Medicine, and Technology: The Difference Feminism Has Made*, edited by Angela N. H. Creager, Elizabeth Lunbeck, and Londa Schiebinger. Chicago: University of Chicago Press.

Mathis, Betty Ann and Harold F. Mathis. 1972. "Women Enrolled in Engineering Curricula." Penn. State University Archives.

Mattfeld 1965. Notes of Jacquelyn Mattfeld; AC134, box 1, folder "Academic council 6/64–6/65." MIT Archives.

McNatt, Bob. 1952. "Petite Blonde is First Tech Coed Candidate." *Atlanta Journal and Constitution*, 13 April, 1.

Michigan State News. 1964. "Female Scientist Image Blasted," *Michigan State News*, November 4, 1964: 1.

Miller, Joy. 1964. "Women Engineers: They're Feminine and So Bright." *Perth Amboy NJ News*, 30 July, 30.

MIT. 1941. *Massachusetts Institute of Technology Handbook*, 1941: 5.

――― 1999. "A Study on the Status of Women Faculty in Science at MIT." http:// web.mit.edu/fnl/women/women.html

MIT Ad Hoc Committee. 1972. *Report of MIT Ad Hoc Committee on the Role of Women at MIT*, ca. 1972, 3. MIT Archives.

MIT Association of Women Students. 1977. MIT AWS notes; MIT Archives.

National Science Foundation (NSF). 2000. *Women, Minorities, and Persons with Disabilities in Science and Engineering*. National Science Foundation Report 2000. http://www.nsf.gov/sbe/srs/nsf00327/access/toc.htm

O'Bannon, Helen. 1975. "The Social Scene: Isolation and Frustration," box 128, file "Women in Engineering—Beyond Recruitment Conference Proceedings", June 22–25. SWE collection, Wayne State University Archives.

Ogilvie, Marilyn Bailey. 1986. *Women in Science*. Cambridge: MIT Press.

Oldenziel, Ruth. 2000. "Multiple Entry Visas: Gender and Engineering in the U.S., 1879–1945." In *Crossing Boundaries, Building Bridges: Comparing the History of Women Engineers, 1870s–1990s*, edited by Annie Canel, Ruth Oldenziel, Karin Zachmann, 11–50. Harwood Academic Publishers.

—— 1999. *Making Technology Masculine: Men, Women, and Modern Machines, 1870–1945*. Amsterdam: Amsterdam University Press.

Peden, Irene Carswell. 1965. "Women in Engineering Careers." Booklet, SWE collection, Wayne State University Archives.

Purcell, Carroll. 1979. "Toys, Technology, and Sex Roles in America, 1920–1940." In *Dynamos and Virgins Revisited: Women and Technological Change in History*, edited by Martha Moore Trescott: Metuchen, NJ.

Rensselaer Polytechnic. 1943. "Curtiss Wright Women Enter Rensselaer To Begin Ten Month Aeronautics Course." February 22: 1.

Rossiter, Margaret W. 1995. *Women Scientists in America: Before Affirmative Action, 1940–1972*. Baltimore: Johns Hopkins University Press.

—— 1982. *Women Scientists in America: Struggles and Strategies to 1940*. Baltimore: Johns Hopkins University Press.

Rutherford, Alta. 1954. "Women Engineers in Redlands Spotlight," *Detroit News*, April 19, 27.

Sloan, Sarah. 1979. "Terry's Trip." *SWE Newsletter*, November/December, 2

Stiles 1946. Florence Stiles to Carroll Webber Jr., March 28, 1946; AC220, box 2, folder 2; MIT Archives.

SWE 1953. "Society of Women Engineers brochure, ca. 1953." SWE collection, Wayne State University Archives.

—— 1993. *Society of Women Engineers Report*, 1993; http://www.swe.org/SWE/ProgDev/stat/stathome.html

SWE Berkeley. 1980. Student Section—University of California, Berkeley; Junior High School Outreach: A Practical Guide, 1980; box 118, file "Junior High School Outreach 1980." SWE collection, Wayne State University Archives.

SWE Newsletter. 1961. "81% of Male Bosses Won't Hire Gal Engineers: Remainder Take Dim View of Middle Management Spots." 7 (March):1.

—— 1978. *SWE Newsletter*, May 1978: 5.

—— 1979. "Entry and Turnover in Employment," *SWE Newsletter*, March–April 1979: 1.

SWE Report on Employment: http://www.societyofwomenengineers.org.

Terry's Trip n. d. ca. 1979. Box #131, file "Terry's Trip." SWE collection, Wayne State University Archives.

Time. 1963. "Where the Brains Are." 18 October 18, 51.

Trescott, Martha Moore. 1990. "Women in the Intellectual Development of Engineering: A Study in Persistence and Systems Thought." In *Women of Science:*

Righting the Record, edited by G. Kass-Simon and Patricia Farnes, 147–187. Bloomington: Indiana University Press.

Van Leer 1948a. Blake R. Van Leer to Sandy Beaver, Nov. 8, 1948; box 2, folder "9", 86-01-08, Georgia Tech Archives.

Van Leer 1948b. Blake R. Van Leer to Harmon Caldwell, Dec. 28, 1948; box 2, folder "9", 86-01-08, Georgia Tech Archives.

Wajcman, Judy. 1991. *Feminism Confronts Technology*. University Park: Pennsylvania State University Press.

Walker, Eric. 1955. "Women are NOT For Engineering." *Penn State Engineer*, May: 18

War Training Programs, 1945. War Training Programs—World War II: Curtiss-Wright Engineering Cadette Training Program, report Program Series A, Iowa State College, v. A I. April 1 1945.

Wick 1970. Emily L. Wick, "Proposal for a New Policy for Admission of Women Undergraduate Students at MIT," March 9, 1970; MIT Archives.

Wick 1971. Emily Wick to Paul Gray, November 16, 1971; MC485, box 13, file "MIT"; MIT Archives.

Widnall, Sheila. 1976. Interview by Shirlee Shirkow. MC86, Box 8, MIT Archives.

Woloch, Nancy. 1999. *Women and the American Experience*. New York: McGraw Hill.

Women in Science and Technology: A Report on the Workshop on Women in Science and Technology. 1973: 2–4; MIT Archives.

3

2000 *Multiple-Entry Visas*

Gender and Engineering in the U.S., 1870–1945[1]

Ruth Oldenziel

IN 1978, ALMOST [A] HUNDRED YEARS AFTER THE FIRST woman had graduated from Iowa State University's engineering program, Samuel Florman wondered in *Harper's* magazine why women failed to take the same existential pleasure in engineering that he advocated for men.[2] Florman, a spokesman for engineers, claimed to have found the answer when visiting the lofty halls of Smith College, a private women's college with an elite reputation. He could imagine the smart women students—all well trained in math and the sciences—'donning white coats and conducting experiments in quiet laboratories.' But he could not see these sensible, bright young women becoming engineers. '[I]t is "beneath" them to do so,' he said. 'It is a matter of class.'[3] He believed that, given the option, Smith women with strong science and mathematical abilities would choose the sciences rather than the low-status profession of engineering. Florman then made a call to feminists to solve the problem of status in engineering by entering the profession.

Oldenziel, Ruth. (2000). "Multiple entry visas: Gender and engineering in the U.S., 1879–1945." *Crossing boundaries, building bridges: Comparing the history of women engineers 1870s–1990s*, Annie Canel, Ruth Oldenziel, Karin Zachmann, eds., Harwood Academic Publishers, Amsterdam, 11–50. Reprinted with permission of Taylor and Francis Publishers.

Unfair, to say the least! But his observation pushes us to take a closer look at the importance of class relations in women's entry into engineering.

In the U.S., engineering changed from an elite profession to a mass occupation, grew the fastest of all professions, and developed specialist areas at a great pace after the 1890s. American engineering was also a deeply divided and segmented profession. Engineers could be found working anywhere from board rooms to drafting departments, mechanics workshops, and chemical labs. They worked as executives, managers, designers, draftsmen, detailers, checkers, tracers, and testing technicians. American engineering did not become a closed profession associated with science as it did in France, where the state groomed a small elite for leadership positions. Nor did it resemble British engineering culture with its small firms, craft traditions, working-class associations, and kinship networks. Instead, American engineering evolved into something between the French and the British models: a mass middle-class occupation with a hybrid form of professionalism and an almost automatic distaste for blue-collar union organization.[4] Compared to the profession in France and Russia, engineering in America was a relatively open affair: doors were opened early to newcomers from the lower classes and different ethnic backgrounds to staff rapid industrialization. There was neither a central agency nor professional organization to certify engineers, nor a national educational system. American engineers were trained either informally on the work floor or formally at a variety of engineering institutions.[5]

The openness of America's engineering system had its limits, however. Engineering advocates in the U.S. were eager to maintain professional prestige, which proved fragile. Here gender and class were linked: even though in the United States the profession was divided and diverse, professional engineering organizations and educational institutions often used sex discrimination as a means by which to preserve class distinctions. At the same time, the fact that the profession lacked central organization nevertheless opened a small window of opportunity for some women. Both engineering educators and the state mobilized women trained in the sciences for their own purposes in times of war (cold and hot). As they faced changing coalitions of men, and old-fashioned hiring and promotional practices, women engineers developed many individual and collective strategies to force their way into the profession.

Bridges of Sisterhood: Shaky at Best

Few women engineers publicly rallied to the feminist movement or its causes.[6] This disinterest was mutual.[7] The American women's movement focused on the sciences rather than on engineering because the latter lacked cultural authority. Unlike their Russian, Austrian, and French sisters, American philanthropists and advocates of women's education neither paid any special attention to the engineering profession as a vehicle for women's equality, nor helped to establish separate engineering institutions.[8] Although in the United States, it was rapidly becoming a mass occupation and the few women in engineering who entered came from a higher class background than their male counterparts, engineering failed to attract the many young elite women looking for a suitable vocation. Raised on the propaganda of the First World War, when the government desperately tried to recruit more technical personnel, a generation of academically-trained women engineers grew up in an era when suffrage had been won and when professional women seemed to make headway. They adhered to a belief that professional status and advancement were based on merit, having nothing to do with gender.

Among American women engineers, only Nora Stanton Blatch (1883–1971) envisioned as mutually enriching the two sides of her life as an engineer and a feminist. As a third-generation suffragist, Nora Blatch could see herself in a feminist genealogy. She descended from a line of famous activists (her mother was Harriot S. Blatch, and her grandmother was Elizabeth Cady Stanton) and campaigned for suffrage at Cornell University, where she had chosen civil engineering as her major: because, she said, it was the most male-dominated field she could find. Her generation of women engineers grew up in the nineteenth century, when the bond of solidarity among women was more firmly entrenched, but Blatch went a step further than her contemporaries. She accused the American Society of Civil Engineers (ASCE) of sex discrimination when in 1916 it tried to bar her from full membership; moreover, she campaigned for pay equity between men and women through the National Woman's Party for many years. Where Blatch found literary and organizational models for professional women in the feminist movements of her mother and grandmother, she found none among her women colleagues in engineering.[9]

Most American women engineers ignored any link between the women in the technical field and those working in the woman's movement. They honored instead an alternative model offered by Lillian Gilbreth (1878–1972). Gilbreth borrowed Rudyard Kipling's rewording of the biblical allegory of Martha ('simple service simply given') as the inspiration for women engineers. Gilbreth, holder of a Ph.D. in psychology, had acquired her technical knowledge and her legitimacy in engineering through her husband—a 'borrowed identity,' as Margot Fuchs terms it, that she expertly managed. She was actually a widow for much of her working life, but she projected an image of herself as a married career woman.[10] Avoiding the open confrontation that

Nora Stanton Blatch, civil engineer and descendant of two generations of women right activists on suffragist campaign on horseback in New York State in 1913, challenged the engineering establishment in 1916 on charges of discrimination.
Courtesy of Coline Jenkins. Used with permission.

Blatch chose, Gilbreth advocated a professional strategy for women engineers based on hard work, self-reliance, and stoicism. *[For another perspective on Lillian Gilbreth, see Chapter 5. Ed.]*

Such a strategy amounted to what Margaret Rossiter described as 'the classic tactics of assimilation required of those seeking acceptance in a hostile and competitive atmosphere, the kind of atmosphere women heading for bastions of men's work encountered at every turn.'[11] These were: 'quiet but deliberate over-qualification, personal modesty, strong self-discipline, and infinite stoicism.' Indeed, American women engineers maintained their loyalty to male models of the profession at great personal cost. Vera Jones MacKay, a chemical engineer who had managed to find work on pilot plants for fertilizer production with the Tennessee Valley Authority, recalled painful memories when looking back on her career in 1975: 'it is hard to discuss my working days as an engineer without sounding like one of the most militant of the women's libbers.'[12] Jones's public admission of the personal costs involved in her career choice is unusual because most women engineers kept a stiff upper lip in the constant struggles before affirmative action. Most women engineers preferred Lillian M. Gilbreth as their role model to her contemporary Nora Blatch. Hence, women engineers drew their literary models and organizational forms from their engineering fathers rather than from their feminist sisters. Before the Second World War, American women engineers—unlike their British counterparts—cultivated silence as a survival strategy and ventroloquized [sic] discontent without ever directly articulating it. As rank-and-file members of the profession working for corporate and military establishments, American women engineers not only became invisible to themselves as a group, but also to history, not least because of the failure to build bridges with women of the women's movement or those of the Progressive movement who helped shape the infrastructures in cities.

Surrogate Sons and the Family Job

A great many American women found their way into engineering through what might be called patrimonial patronage and matrimonial sponsorship. Most of them never appeared in any statistics because their ties to engineering were through their fathers, brothers, and husbands. Supported by kinship ties, such familial patronage and sponsorship often offered relatively easy access but also resulted in what we have already seen in the case of Lillian Gilbreth: a borrowed identity. Most women in such circumstances did not have female role models but looked to their fathers and brothers. Their identity as engineers was therefore largely 'on loan,' even if some, like Gilbreth, managed to stretch the terms of the loan to the limits of social approval.[13]

Formal training was an important credential for continental European countries such as France and served as a wedge into other professions in the U.S. But it did not play such a decisive role in employment opportunities in American engineering before 1945. By the end of the forties and beginning of the fifties, only fifty-five percent of American men—and even fewer women (twenty percent)—who worked in engineering had completed such training. In the nineteenth century and probably well into the twentieth century, a few hundred women continued to manage their late husbands' engineering work, having received enough informal technical training to call themselves engineers.[14] Others besides Gilbreth became well-known after learning their trade through family and husbands. They acquired technical knowledge on the job or through an informal system of education within

family firms without ever attending a specialized school. Emily Warren Roebling (1841–1902), who kept the family firm going when illness kept her
husband housebound, acted as his proxy throughout most of the building of
New York's Brooklyn Bridge in the 1870s and 1880s. Trained in mathematics, she learned to speak the language of engineers, made daily on-site inspections, dealt with contractors and materials suppliers, handled the technical
correspondence, and negotiated the political frictions that inevitably arose
in such a grand public project. The Brooklyn Bridge had been a Roebling
project on which the family's fortunes depended, and Emily Roebling served
as her husband's proxy for decades. [For more on Roebling, see Chapter 6.
Ed.] Most wives, however, worked anonymously in family businesses. As
late as 1922, a woman active in civil engineering wrote to the editor of The
Professional Engineer saying that she greatly appreciated that the journal
finally acknowledged the wives of engineers without specialized degrees:
'My training in engineering began with marriage and I have filled about
every job . . . from rodding and driving stakes to running a level party, or
setting grade and figuring yardage in the office.'[15]

Most such women would be forgotten. Lillian Gilbreth, however,
became America's most celebrated woman engineer, in part because she
used her borrowed identity to do the widest possible range of work while
still appearing to conform to social norms. Frank Gilbreth's untimely death
in 1924 must have been devastating for a mother with twelve children but,
aided by a team of domestic hands, it also allowed Lillian Gilbreth to enjoy
considerable freedom in her role as a widow for nearly fifty years. She
expertly managed her image, fostering publicity that cast her in the role of
a married career woman. This public persona provided perfect protection
against possible disapproval of her career ambitions. Similarly, long after
Frank Gilbreth's death, she allowed her marriage to be publicized as a partnership to be emulated: the perfect, most efficient match between business
and love. Such a partnership was analogous to her husband's advocacy of
performing tasks in the 'one-best-way' for maximum benefit.[16] As Gilbreth's
strategy shows, family connections might guarantee work, and they could
also serve to protect women engineers from public scrutiny. Newspaper
reports and government propaganda played up women engineers' strong
family ties to men as a way to ward off any possible threat from these female
incursions into the male domain.

The effort to 'domesticate' women's talents into familiar categories
prevailed during the Second World War, when war propaganda emphasized
women's family ties to engineering. Most of the available biographical information on the social background of women engineers was generated as part
of this campaign to attract women engineers. Historical narrative sources

on women engineers therefore tend to overexpose women with family connections. Nevertheless, it is clear that formal engineering education, with or without a degree in hand, could be useful for some, such as the daughters of proprietors of small manufacturing firms. For example, Beatrice Hicks trained at the Stevens Institute of Technology, going on to become chief engineer, then vice-president, and finally owner of her father's Newark Controls after his death. After her graduation in engineering Jean Horning Marburg supervised the plant construction for her family's mining property in Alaska. Florence Kimball was another graduate who worked at her family's elevator firm, drafted plans for the remodeling and building of its real estate property, and drew several blueprints for patents—the most exacting of all draftsmanship. Small family firms like the Kimball and Horning companies not only tried to maximize production and profits, but were also in the business of building and maintaining a family legacy.[17]

Succession in patriarchally organized family firms was exclusively an affair between fathers and sons, but circumstances sometimes pushed daughters into becoming surrogate sons. The most celebrated and best-documented case is that of Kate Gleason (1865–1933), the oldest daughter of William Gleason. He had started his own tool making shop, the Gleason Gear Planer Company in Rochester, New York, which later became one of the largest of its kind. Encouraged by the example of early feminist Susan B. Anthony, and prompted by the early death of her half-brother, Kate Gleason began to take courses in mechanical engineering at Cornell University and the Mechanics Institute in Rochester. Her training followed the course of many sons of other family manufacturing firms, who no longer were expected to master a craft completely but had to have a working knowledge of all the various aspects of the firm. Gleason was her family firm's business manager for many years while the business grew dramatically, becoming a major player in the industry.[18] *[For more on Gleason, see Chapter 7. Ed.]* Patrimonial patronage thus encouraged daughters like Kate Gleason to seek formal education with or without completing a degree because it fit into a family business's strategy. The link between business sense and family interests could make such education and work acceptable.

For similar reasons husbands encouraged their wives to seek formal training. Such matrimonial sponsorship not only gave legitimacy to women's engineering accomplishments, but also offered them the hope of establishing a firm in partnership with their husbands. The pooling of resources of man and wife in an enterprise afforded the opportunity for a partnership of business and love. Sometimes, however, engineering marriages could turn into a liability. Many women met their partners at college or in the field of engineering, allowing them to enter into male social and study circles otherwise closed to them. But because of the inherent power inequality in such mentorship, such relationships could turn into a distinct disadvantage for the wife's career advancement later if she questioned the terms of her borrowed identity or the matrimonial sponsorship.[19]

As a young feminist activist and engineering graduate Nora Blatch and her husband, the engineer and inventor Lee De Forest, shared in the excitement of new emerging technologies such as the radio at first. But in the end they disagreed about who was to shape and direct the possibilities of these developments. On their first meeting, Blatch 'tremendously admired' the young radio inventor Lee De Forest and recalled that 'a life in the midst of invention appealed to me strongly.'[20] For his part, De Forest thought 'destiny' had brought her to his door and pursued her relentlessly. In desperate need of money for various ventures, he accepted funds from his future mother-in-law Harriot S. Blatch, while Nora's technical training, her love for music, and the connections with the New York powerful, brought enormous technical, financial and social resources to his flagging career. No doubt seeing an opportunity to fulfill her life's goal of combining career and marriage, Nora Blatch took extra courses in electricity and mathematics with Michael Pupin, a well-known New York electrical engineer, and worked in De Forest's laboratory on the development of the radio. Together Lee and Nora were able to air the first broadcasts of music and conversations in the New York area. On their honeymoon to Paris, the newlyweds seized the opportunity to promote their wireless phone by a demonstration from the Eiffel Tower, organized through Blatch's family connections.[21]

Both Blatch and De Forest shared an excitement about participating in the new technological developments with their contemporaries. To Harriot S. Blatch (and no doubt to her daughter as well) technologies such as the radio were new tools for women to use for their own ends. At one of the promotional experiments for the 'wireless phones' in New York in 1909, Harriot, Nora, and Lee were positioned at the Terminal Building, while a group of women students from Barnard, their physics professor, and some male interlopers from Columbia stood listening to the transmission at the Metropolitan Life Building. 'I stand for the achievements of the twentieth century,' Harriot Blatch declared in the first message sent. 'I believe in its scientific developments, in its political development. I will not refuse to use the tools which progress places at my command . . . not forgetting that highly developed method of registering my political opinions, the ballot box.' Since the transmitter was only a one-way communication, she continued uninterrupted— although a male student from Columbia protested that 'that is a mean way to talk at a poor chap when he can't say anything.' Believing that technological modernity was inextricably and inevitably linked with politically

progressive ideas, she continued: 'Travel by stagecoach is out of date. Kings
are out of date: communication by canalboat is out of date; an aristocracy
is out of date, none more so than a male aristocracy.'[22] The speech was used
by De Forest and his business agent to sell stock of his Radio-Telephone
company to suffragists and their supporters.[23] Even if Nora Blatch and Lee
De Forest shared in the excitement of the new technologies, disagreements
emerged over the financial status of the firm once they had married and their
child was born. De Forest opposed his wife's management opinions and her
insistence on continuing to work in engineering. This caused their separa-
tion in 1911. Explaining his divorce, De Forest told reporters of a national
newspaper that 'his matrimonial catastrophe was due to the fact that his
wife . . . had persisted in following her career as a hydraulic engineer and an
agitator [for women's suffrage] . . . after the birth of her child.'[24] He warned
other men against employing their wives, conveniently omitting all mention
of Blatch's technical and financial participation in his ventures. Eventually,
Blatch started her own architectural firm with family capital. It allowed her
to remain independent from partners like De Forest and from the corporate
employers she had earlier learned to avoid. De Forest and other husbands
were interested in joint ventures but not an equitable partnership with their
wives. De Forest, who insisted that he wholeheartedly supported suffrage
for women, had admired Nora's intelligence, and employed her technical
training. Yet his wife's greatest offense was that after marriage and motherhood she had rejected a loaned identity and continued to assert her feminist
beliefs and heritage.

School Culture and the Strategy of Overqualification

Family businesses were based on a form of engineering knowledge which
linked them to the patriarchal authority of the traditional workplace.
Formal education, by contrast, was to be a more democratic form of knowledge accessible to all, but it was still in need of establishing and reproducing
its own male model of authority. In the decade following the American Civil
War, diversity and openness characterized American engineering education,
but nevertheless it also used sex and race in exclusionary ways. Hailed as
the landmark legislation that pushed co-education to unprecedented levels,
the Land-Grant Morrill Act of 1862 helped establish several schools of
engineering at land-grant state universities, colleges, polytechnic institutes,
and private universities throughout the land. Its drafters had intended it

for education of the children of farmers and industrial workers, but had not stipulated the character of 'agricultural and mechanic arts.' At places such as MIT, women, workers, and farmers attended courses in the early days. Industrialists had been the first to support education of workers and women, viewing them as potentially well-disciplined work force.[25]

This broad interpretation of the Act changed over the course of the century. 'The agricultural and mechanic arts' often came to mean industrial rather than agricultural education, technical rather than agricultural training, and school-based engineering rather than a British-style apprenticeship. Engineering educators began breaking with the traditions of vocational training, latching onto scientific rhetoric. The push to upgrade the field through professional ideals resulted in the masculinization of higher engineering education, sending women into separate fields of chemical lab work or home economics.

Before this closure, American women were welcomed as special tuition-paying students when engineering educators sought to increase their enrollment figures in new programs. Compared to their sisters in other countries, American women had free access to primary and secondary education and came relatively well prepared. The newer institutions in the US were more welcoming than the established ones. Thus the co-educational land-grant institutions and state universities showed a more favorable attitude towards women's higher education in engineering than privately owned and sex-segregated institutions such as denominational colleges, military academies, and high-status private schools. The state sponsored land-grant institutions (e.g., Purdue, MIT, Iowa State, Ohio State, Cornell, Berkeley, and the Universities of Washington, Illinois, Colorado, Michigan, and Kentucky) and many municipal universities (the Universities of Cincinnati, Louisville, New York, Houston, and Toledo) thus led the way in engineering education for women. Even some mining schools admitted women to their engineering departments.[26]

In the pre-professional era, engineering institutions and occupational clubs had not yet achieved prestige comparable to that in other professions. Nor had home economics yet been established as a separate field for women interested in technical fields and applied sciences. Pioneering women students therefore began to graduate in engineering from the 1870s onwards. Even so they received mixed messages. Engineering educators searching for higher enrollments might admit some women to their programs, but women faced outright discrimination at every turn, and staying the course required tremendous stamina. A complete set of data on the enrollment and graduation figures of three schools (Ohio State University, University of Alabama, and Stanford University) suggests—not surprisingly perhaps—that the dropout rate for women was 25 percent higher than for men, 50 instead of 40 percent of those enrolled.[27] Even women who managed to complete their course work did not always receive the official recognition they deserved. The experiences of Lena Haas at Columbia, Eva Hirdler at the University of Missouri (1911), and Mary and Sophie Hutson at Texas A&M (1903), are telling examples of women students who satisfied all their requirements without receiving the appropriate degrees. Engineering educators were trying to raise academic standards to compete with colleagues in the humanities by playing down the achievements of their women recruits who boosted enrollment figures for their newly established programs.[28]

Facing discrimination, women engineers paired stoicism with a strategy of overqualification. The experienced mechanical engineer Margaret Ingels warned in the 1930s that a woman engineer 'must in many cases work even harder than a man to build up confidence.' *[Ingels is profiled in Chapter 10. Ed.]* Two decades later another woman found the situation unchanged and concluded that 'a dedicated woman can succeed [but has to] run twice as hard as a man just to stay even.'[29] Women who were willing to fit into the tight-knit male world of engineering could force the doors slightly further ajar by concentrating on their math abilities and doubling their efforts. Many women opted for multiple degrees.

If women engineering students in the nineteenth and early twentieth centuries faced formidable difficulties, lack of preparation in mathematics does not seem to have been one of them. In high school, for one, American girls and boys received an equal amount of instruction in calculus and geometry.[30] Moreover, because women who entered engineering tended to come from higher social strata than their male counterparts, they often had a better general education. In the 1970s, sociologist Sally Hacker argued that the high standards of math in engineering education effectively served to exclude women. But at the point where professional standards were just beginning to be formed, math offered a brief window of opportunity for those women interested in a technical education. Before the Second World War an understanding of mathematics was required for practicing engineering, but in America it did not form the kind of obstacle or rite of passage that it would become later as educators sought to raise the standard of engineering. On the contrary, many women who went into engineering could claim superior ability and knowledge in mathematics. The increased importance that engineering educators placed on mathematics as a means of upgrading the profession might have been a major hurdle to many engineering students with average ability—both women and men—but it also helped brilliant women in a school culture that stressed academic skills over hands-on experience. Exceptionally competent women like Elsie Eaves (1898–1983),

Alice Goff (b. 1894), Dorothy Hanchett (1896–1948), and Edith Clarke (1883–1959) used their mathematical skills and multiple degrees as a wedge into engineering work and mobilized them as a shield against outright discrimination. *[Eaves is profiled in Chapter 10, Clarke in Chapter 9. Ed.]*

Educational reformers such as Robert H. Thurston who sought to upgrade engineering training with a new emphasis on mathematics, history, and the humanities faced a dilemma. Their form of engineering knowledge was not linked to the patriarchal and class authority of the workplace, but was based on the new cultural authority of science and math. Not only were academic engineers often accused of failing to prepare their students to face the reality of the production floor, but academic ideals threatened to become associated with gentility and femininity. In balancing these elements, engineering educators became the most articulate purveyors of an academic male ethos that stressed hands-on experience and a slap-on-the-back kind of manliness. Many engineering educators tried to imitate 'the methods and manners of real shop-life' in college workshops that housed steam engines, blacksmith tools, foundries and the like. Here hands-on experience could be acquired in association with academic ideals. The problem with the college shops, however well-equipped, was that it was impossible to simulate or test a confrontation with the attitudes of independent workers and bullying foreman. The ability to 'handle men'—to be a professional manager—remained the true hallmark of the successful engineer. In the schools of engineering, this managerial ideal balanced precariously between working-class manliness and academic gentility.

In these environments, women students were encouraged to take math classes but often excluded from taking shop classes or field trips to factories that were required for graduation. When Nora Blatch's classmates prepared for a photograph showing them at work in the field, they arranged for a male friend to take Nora on a date on the day of the photo session so that she would not be in the picture. They thus deliberately excluded her from the rite of passage and erased her from the visual historical record. In 1925, MIT professors prohibited Olga Soroka from participating in a field trip required for graduation in civil engineering. Her professors organized a special internship with the New York subway system for her instead, considering it more appropriate to then-current ideals of women's public behavior. Anna Lay Turner, a chemical engineering student at Rice University in 1924, recalled that women were tolerated in academic environments, but barred from mechanical laboratories. Simply to don overalls was to challenge prevailing social codes.[31]

The workshop and building sites thus functioned as a way of screening out women. '[For] it must be clearly understood,' as one critic of their presence in engineering and other technical occupations wrote in 1908, that 'the road to the drafting board and the laboratory of the engineer lies through the workshop, and workshop practice means hard work and blistered hands, not dilettante pottering and observation.'[32] Women might be competent in drafting, calculation, research, and analysis, as employers attested in the 1920s, but sweat, dirt, and calluses made the engineer a real man. Or, in the words of one scholar, 'if science wears a white lab coat, technology wears a hard hat and has slightly dirty fingernails.'[33] Ideally, middle-class men belonged on the production floors and building sites where they managed other men, while women dealt with more technical details in respectable environments. But of course this was only an ideal. Had it been achievable, there would have been no need constantly to reassert its importance.

Foot Soldiers of Bureaucracy

Ever since the First World War, most women engineers were employed by the emerging military-industrial complex. Women also found their way into engineering through corporate and federal apprenticeship, particularly when the government and the private sector worried about the shortage of technical personnel in times of war and competition with foreign countries. Among entry-level jobs, women made the most headway in the laboratory-oriented and newer fields, which required more academic skills and where gender coding was not yet fixed: chemical analysis, electrical, and—after the Second World War—aeronautical engineering. If small businesses provided a way into engineering for women with family ties, the large emerging corporations did so for women without family capital or resources, to whom mechanical engineering—which was steeped in craft traditions—remained a closed shop. This is not to say that women could not be found working in mechanics shops: during the First and Second World Wars, corporations hired working-class women as lathe and punch press operators and as assembly workers. Mechanical engineering implied a different level of work than the engineering trades, however, one that involved supervision.[34] These encouragements of women's engineering work were all temporary, relatively low level, and deeply ambivalent.

For employees of large corporations without family connections or capital, an engineering job held the promise of promotion, even if this became more a vision than a reality over the course of the twentieth century. Formal and bureaucratic rules made gender discrimination endemic. But they also helped to secure better opportunities for women engineers than the informal rituals of firms, whose shop-floor culture encouraged male patterns of advancement. Only a few women, such as Kate Gleason, could crack the male code of the shop floor by invoking an authority that stemmed from family ownership. The growing importance of formal rules and the move toward professionalization in the twentieth century therefore proved to be a double-edged sword for women engineers who entered the profession without capital or connections. The two world wars—and the state—offered opportunities, but not full-fledged careers, while the war economy institutionalized old discriminatory practices and created new ones.

Women might have found formal education a viable means of access to entry-level engineering jobs, but those who excelled in the academic setting did not fare well in their subsequent careers. Highly trained women including Elsie Eaves, Olive Dennis, Patricia Stockum, and Mabel Macferren, all of whom had earned two or three degrees and showed the stamina to succeed, found that this initial advantage turned into a liability once they entered the workplace. In the new environment, male codes of managerial command and hands-on experience determined one's professional standing, rather than academic excellence. Many overqualified women ended up either as high school teachers in mathematics and sciences or as calculators in corporate offices and at research institutions. Dorothy Tilden Hanchett first trained in civil engineering at the University of Michigan ('17). No doubt she believed that gaining additional M.A. and Ph.D. degrees at Columbia University ('27) and Logan College ('45) would help advance her career. Instead, she ended up at Battle Creek High School as head of the math department. In the aftermath of the First World War, Hanchett and many other highly qualified women found that government propaganda had been empty rhetoric. They were forced to accept temporary teaching jobs in elementary and high schools, teaching instructorships at engineering colleges, or editing positions in professional organizations.[35] The tactic of obtaining multiple degrees did not guarantee employment.

In times of economic downturn, only government bureaucracy and highway projects could offer employment (badly paid) to academically trained women. Proportionally more women trained in civil engineering than any other engineering specialization. But in this already overcrowded labor market, they earned the lowest salaries, ending up in low-level positions and finding fewer employment opportunities than in any other branch of engineering. Still, while these jobs might have been demeaning for young men who expected management positions, for women they offered relatively high wages compared with other jobs available to them. The drafting departments of the State Highway Commissions gave temporary jobs to Esther Knudsen in Wisconsin and Elsie Eaves in Colorado during the 1920s, to Myra Cederquist in Ohio in the 1930s, and to Emma Crabtree in Nevada in the 1940s.[36] The large corporations such as Westinghouse, General Electric, and Boeing also offered women an avenue to technical training through a kind of corporate apprenticeship. At Westinghouse, Bertha Lamme found ample opportunity to use her superior mathematical knowledge and her engineering skills to design motors and generators for over ten years, until she had to relinquish her job when she married a co-worker in 1905. *[Lamme is profiled in Chapter 8. Ed.]* Finding the door to engineering slightly ajar during the war in 1917 as a young civil engineering graduate from the University of Michigan, Hazel Irene Quick established a long career that lasted until 1950. As a fundamental plan engineer, she was in fact the only woman employed by the Michigan State Telephone Company. Nevertheless these were mostly individual women who advanced through the ranks at a much slower rate than their male colleagues.[37]

Even if the World Wars offered opportunities to women, employers also responded to the modest increase in the number of women by setting up clear boundaries against women and by creating separate social and spatial arrangements for their female employees. To deal with the small increase of women, employers instituted sex-segregated offices and drafting departments where some academically trained women could move into supervisory but temporary positions. After graduating from Iowa State University in civil engineering in 1894 and doing some graduate work at MIT, Alda Wilson (b. 1873) worked in architectural firms in Chicago and New York for over ten years before she found a managerial job—as superintendent of the women's drafting department at the Iowa Highway Commission in 1919. Unable to find an engineering position after the First World War, the overqualified and brilliant Edith Clarke spent several years training and supervising women in the calculation of mechanical stresses in turbines. This was in a separate women's department that existed within the Turbine Engineering Department at General Electric in Schenectady, N.Y. Such separate female spaces might have offered women a temporary niche but rarely a solid stepping stone for full-fledged careers as designers, executives, or managers.[38]

Before the era of affirmative action, neither the federal government nor the corporations offered true alternatives to the kind of patriarchal patronage experienced by daughters in family firms like Gleason. At the end of her career in 1947 when the government campaigned for women's return to the home, the experienced Olive Dennis still believed in meritocracy. '[W]e certainly do not want to discourage the ambitious young woman with the right qualifications for an engineering career,' she said, but she warned, 'anyone pioneering in this field must be made to see that, outside of the lowest levels of clerical and manual work, there are almost no standard [management] jobs for women.'[39] *[Dennis is profiled in Chapter 10. Ed.]* Dennis knew what she was talking about. A whole generation of qualified women engineers and scientists had weathered the storm during the Depression in order to continue their careers. As Nora Blatch observed when she worked as an engineering inspector for the Public Works Administrations in Connecticut and Rhode Island, federal sponsorships of women were limited during the New Deal. The investments of Roosevelt's public-work administration in major building programs provided engineering work for men only. The reforestation, highway, building, and reclamation projects were all closed to qualified women while they mobilized men for lower wages. (Ironically these men then had to travel far away to their work, leaving their women as heads of household.) Moreover, the National Recovery Board still specified lower minimum wages for women than men.[40]

These highly educated women of the post-war era waited for better times and looked for jobs in teaching, drafting, or editing and secretarial positions with engineering firms and professional organizations. Elsie Eaves, for example, located herself strategically as manager of the business news department of the journal, *Engineering News-Record*. A graduate of the University of Colorado ('20) and from a prominent family, she provided mentoring and career guidance to many young women engineers during the 1930s and 1940s. She counseled them on how to get through the Depression, and encouraged them to acquire stenographic and secretarial skills in the hope of 'a position with a fine engineer.' But she warned, 'I never encourage a girl to study engineering on the theory that if she wants it badly enough she will do it in spite of all discouragement.'[41] During the Second World War, when younger men went to the front and others moved up to fill their positions, women stood ready to take on the new jobs that were opening up. But these never materialized. Instead of recruiting among the experienced women already available, the federal government chose to train new young women still in college. The state thus helped to institutionalize old patterns, best-illustrated by the gender coding of the federal job of engineering aide.[42] This was used during the Second World War to define women as non-engineers.

American women, like those in Europe, were encouraged for the first time to seek training in technical work for the war effort. Under the auspices of the federal government and in cooperation with universities, large corporations urged young and bright women to apply for engineering jobs. Federal agencies, large aircraft companies, and engineering schools pushed over 300,000 women through various kind of engineering programs ranging from three-month courses to college engineering curricula condensed into two years. Georgia Tech University in the South, which like many other well-established schools had been hostile to any hint at coeducation before the war, opened its doors to women for a special training program sponsored by the US Chemical Services when shortages of technically trained personnel threatened the war industries. The aircraft corporation Curtiss-Wright sponsored a course for women engineering students at several American universities including Iowa State, University of Minnesota, and Rensselaer Polytechnic. The women received an engineering certificate after completion of the course which included work in engineering methods, mechanics, drafting, and processing. Many of these specially trained women, who had been the best and the brightest in their high school and college, ended up in drafting, testing, and routine lab work, however. Juliet K. Coyle trained in medical technology and biology as a young college student in 1943, when she was recruited by an aircraft company for a short course in engineering where she learned to read blueprints, drafting, statistics, and mechanical practice. When the war ended and she finished her studies, however, the company had no idea what to do with her and her equally well-trained women colleagues. The firm not only moved them around through different departments and paid them less than their male counterparts, but gave them explicit instructions to avoid giving orders to workers on the production floor—the male avenue to further promotion on the managerial ladder.[43]

Coyle's experience illustrates that none of these educational efforts, either during or after the war-time labor shortages, were meant to turn women into full-fledged engineers. All programs were clearly intended for women's temporary employment as technical assistants to the various engineering departments, despite propaganda claims to the contrary.[44] During the Second World War, vocational literature attempted to assure female recruits that such engineering work would lead to full careers, but Margaret Barnard Pickel, an adviser to graduate students at Columbia, questioned that promise and advised women to prepare for a backlash in peacetime. 'Are the educators of women justified in encouraging their students to start on the long, arduous and expensive training for an engineering degree with the expectation of a career at the end of it?' she asked. After taking an inventory of the barriers women would face, she concluded, 'it seems hardly honest to hold out such a prospect as a professional possibility for women.'[45] Olive Dennis, often quoted by the Baltimore and Ohio Railroad in their public-relations literature during the war, also pointed out that 'Women engineers have been ignored or else glamorized with newspaper publicity that is harmful to serious advancement in their work.'[46] The investigation by the Women's Bureau into employment opportunities in peacetime was equally realistic and cautious. In explaining their research project, the officials at the Women's Bureau stated in their correspondence that 'it is our opinion that the increase of women [in technical and scientific work] during the war

Women trainees in aeronautical engineering posing for propaganda photograph for the Curtiss-Wright Corporation's Cadette Program, an aeronautical engineering crash course for women during the war in 1943.

Courtesy of Iowa State University Library, Special Collections Department, used with permission.

Three women draftsmen trained as engineering aides posing in classic Rosie-the-Riveter government propaganda style in 1943. The photo campaign promoted by the Department of Labor's Woman's Bureau advertised women workers' importance to the war industry.
Courtesy of the Schlesinger Library, Radcliffe Institute, Harvard University, used with permission.

has been greatly exaggerated because of the publicity presented to attract them. However, we want to find the facts through first-hand contact with professional organizations . . . those who employ women in technical and scientific jobs, and with training centers.'[47]

Indeed, Margaret Pickel had been justified in sounding the alarm, as the Women's Bureau found. Strategies to contain women—most of whom were college educated—took various forms. Not only was a woman's job given the title of 'engineering aide,' but an implicit division of labor by sex relegated women to drafting departments and laboratories, while men were assigned jobs on the production floor that enabled them to advance to managerial positions. In short, the federal policy created the term 'engineering aide' to refer to women engineers, while it continued to use the title of 'engineer' to denote men. The job title of 'engineering aide' thus forcefully drew a line between technical expertise and management, and both reproduced and helped to create standard practice.[48]

In the United States, as elsewhere, understanding the formulation of a male professional identity requires attention to the relationship between management and the technical content of engineering as an occupation. Overqualified women found that the initial advantage of academic education would turn into a liability once they entered the workplace, where male codes based on managerial qualities and hands-on experience determined one's professional standing. The gender division of engineering labor—between production floors and drafting offices—hinged on the very same (class) distinctions by which male engineers sought to distinguish themselves from self-trained foremen who had risen through the ranks. The division of labor between the laboratory and the drafting department on the one hand, and building sites and production floors on the other, became the single most important delineator between men and women in engineering. Wherever women engineers did succeed in gaining employment, they were most likely to be hired in drafting, calculating, or design departments, or laboratories and classrooms. In other words, women engineers joined the rank-and-file of the profession without hope of advancement.

No woman without a family connection ever moved into supervisory positions in family firms. Women like Roebling, Gleason, and Gilbreth, who were steeped in the patriarchal culture of the family business, advanced in engineering through a combination of excellence, perseverance, family connections, and the pooling of resources. But assessing the chances of women's employment in 1940, Olive Dennis (1880–1957), who like Blatch had received her education from Cornell University in civil engineering ('20) in addition to degrees in mathematics, warned that 'unless a woman has a family connection in an engineering firm, or enough capital to go into business herself, her chances of rising to an executive position in structural engineering seem negative.'[49] Employed first as a draftsman in the bridge division of the Baltimore & Ohio Railroad and later transferred to the company's service department for interior coach design, Dennis' response during the 1940s is the most revealing: she was always touted as a woman's success story, both by women engineering advocates and by her employers, especially during the Second World War, when the War Manpower Commission and the Office of War Information launched an intense propaganda campaign to lure women into the technical fields. Dennis's warning is instructive in many ways. It pointed to the split between government rhetoric and women's experience of it; the difference between women with family resources and those trying to make it on their own in the emerging corporate and federal bureaucracies; the gap between women's technical expertise and their ability to move into managerial positions; and the contrast between nineteenth-century ideals and twentieth-century practices.[50]

More explicitly than any woman might have said it, the introduction of the term 'engineering aide' encapsulated the story of women's marginalization as a labor reserve force. With a single linguistic stroke the term placed women with technical ability and training outside the domain of technology.

Facing Male Professionalism

National professional organizations became the most visible, if not the only, institution of the engineering fraternity. Few scholars still regard the nineteenth-century creation of professional identities as a trend towards expertise, knowledge, rational behavior, peer review, and non-ideological values. Most now consider professionalization a form of occupational control by which autonomy for some, with carefully defined jurisdictions and privileges, was guaranteed by a rhetorical mask of political disinterestedness and objectivity. The engineering societies were by no means exceptional; they explored some of the classic strategies pioneered by other professions, looking for new ways to enhance their status and cultural authority. The classic model of professionalism was medicine, which emphasized the work ethic, trust, professional associations, licensing, collegial control, and strong client-practitioner relationships. This was problematic for American engineers, however, as many historians of engineering professionalism have argued. In engineering the classic model therefore also competed with what Peter Meiksins identifies as the business and rank-and-file models of professionalism. The American Institute of Mining and Metallurgical Engineers (AIME) advocated business values, associating itself closely with the culture of family firms, and adhering less strictly to the newly emerging ideology of professionalism: profits were more important than ethics. Its business-oriented policy had an immediate impact on the number of women admitted: while only twenty-five women had majored in mining by 1952 and many states had laws prohibiting women from working underground, the mining engineers admitted more women to their ranks than any of the other major engineering organizations. In 1943, the AIME membership included such daughters of family firms like Jean Horning Marburg, then member of the National Resources Planning Board, Helen A. Antonova, an assayer at the R&F Refining Co., Edith P. Meyer, a development engineer at Brush Beryllium Co., and another 19 women in addition to a large number of female students.[51] For daughters of family firms like Jean Horning business professionalism opened some doors that would have remained closed otherwise. Thus business models of professionalism were more open to women if they were connected to the patriarchal culture of family firms like Gleason, Gilbreth, and Horning. They were closed to women without the proper family connections.

Even if engineers did not succeed in maintaining strict boundaries compared to the other professions, when successful the classic model of professionalism ensured a thoroughly male and middle-class endeavor. The more an organization strove to follow the example of medicine, the more it was inclined to bar women.[52] Here it was not the number of women engineers in either absolute or relative terms that determined the percentage of female membership, but rather the level of professionalism to which the leadership of national organizations laid claim. Significantly more women engineering students opted for civil and electrical than other engineering fields, but this was not reflected in the membership of the American Society of Civil Engineers or the American Institute of Electrical Engineers, whose requirements were far more strict than those of AIME.

The American Society of Civil Engineers (ASCE) and other major societies guarded against any female incursions. Some also granted secondary membership without voting rights to unimportant rank-and-file engineers and women who managed to infiltrate them.[53] The American Institute of Electrical Engineers (AIEE), emerging in a field with high aspirations towards a medical model of professionalization, refused to admit Susan B. Leiter, a laboratory assistant at the Testing Bureau in New York, to membership in 1904, when the organization was looking for ways to upgrade the profession.[54] When Elmina Wilson and Nora Blatch applied for membership to the American Society of Civil Engineers, they found the doors closed. Having graduated *cum laude* in civil engineering from Cornell University, Nora Blatch could claim superior mathematical ability and theoretical engineering knowledge, the kind of credentials advocates of engineering schools thought crucial for any engineer to succeed. But Blatch had more to offer: she also possessed the necessary hands-on experience that advocates of shop floor and field training saw as hallmarks of the true engineer. In 1916, however, the society dropped her from its membership and Nora Stanton Blatch filed a lawsuit against the ASCE. As an experienced engineer, she had accumulated over ten years of experience to meet the Society's requirements. In addition to her four-year education in civil engineering at Cornell, she had taken courses in electricity and mathematics with the famous Michael Pupin at Columbia University. She had practiced as a draftsman for the American Bridge Company and the New York City Board of Water for about two years and as an assistant engineer and chief draftsman at the Radley Steel

Construction Company for another three years. Finally she had worked as an assistant engineer at the New York Public Service Commission. Most importantly for the requirements for full membership, she had supervised over thirty draftsmen when working at Radley Steel. Blatch, a feminist, divorcee, and single mother whose income depended on her engineering work at the time, challenged the ASCE when more women and sons of lower-class men were trying to enter the field through the new institutions of higher education and when engineering advocates were busy defining the occupation as a profession by excluding more and more groups of practitioners such as draftsmen and surveyors. Her suit marks one of many contests in which the emerging professions staked out their professional claims by means of border disputes with other competing fields.[55]

In addition to outright exclusion the professional organizations also granted women secondary membership without voting rights. The controversy over Ethel Ricker's Tau Beta Pi membership was yet another drawn-out contest about gender in the fields of civil engineering and architecture, where women were numerous. In 1903, the local chapter of Tau Beta Pi elected Ricker, an architecture student at the University of Illinois, to the engineering honor society. But the national executive board and the society's convention not only overturned the decision to elect her, they went so far as to amend the constitution to specify that henceforth only men would be eligible for membership. During the Depression of the 1930s, when many civil engineers faced unemployment, the Tau Beta Pi honor society introduced a Women's Badge in an attempt to deal with the (small) number of qualified women who had made their presence felt. Women's Badge wearers were neither members nor allowed to pay initiation fees. Disapproving of such 'separate but unequal' recognition, some women refused them. It would take three quarters of a century before women would be accepted as equal partners in the organization: in 1969 the honor society changed its constitution to admit women, in 1973 it cleansed the constitution and bylaws of sexist language, and in 1976 it elected a woman as a national officer for the first time. By changing its constitution and by designing a 'woman's badge,' the fraternity of young aspiring engineers set up explicit gender barriers around engineering when job markets were particularly tight.[56]

As in the various scientific fields, engineering specializations dealt with the perceived threat of feminization either by excluding women from full membership in professional organizations or by relegating them to a secondary status without voting rights. In the words of Margaret Rossiter, 'the very word professional was in some contexts a synonym for an all-masculine and so high-status organization.'[57] Women faced outright exclusion, were relegated to secondary membership, and banished to separate organizations.

Divide and Conquer

Defining separate labor markets for male and female engineering work was another tactic by which to cope with women who started to seek engineering education and employment. The best example comes from the chemical industry, where many women worked as chemists. To deal with these female incursions the American Institute of Chemical Engineers did everything in its power to define the occupation in such a way as to effectively bar women from the field and relegated them to chemistry.

In an effort to distinguish themselves from chemical analysts—a large proportion of whom were women—whose status and pay diminished dramatically around 1900, chemical engineers placed heavy emphasis on the ability to manage other men. In response, production chemists sought to align their occupation with the male world of mechanical engineering rather than with science. Chemical engineers saw themselves running plants, as opposed to laboratories. As one of the most important founders of the American Institute of Chemical Engineers (AIChE), Arthur D. Little (1863–1935), spokesman of engineering professionalism and the nestor of commercial chemical research, introduced a key concept for the development of a distinct chemical engineering identity in 1915. This was the notion of 'unit operations.' Little argued that unit operations involved neither pure chemical science nor mechanical engineering, but distinctly physical, man-made objects in the plant operation rather than chemical reactions in the laboratory. In the same year that Little refined his notion of chemical engineering in a report for the AIChE, his opinion was solicited by the Bureau of Vocational Information, which was preparing a report on employment opportunities for women in chemistry and chemical engineering in the tight engineering labor market after the First World War. Recommending chemical analysis as 'one of the most promising fields of work for women,' Little reserved chemical engineering as an exclusive specialization for men, arguing that 'it is probably the most difficult branch of the profession.' In addition to requiring long hours, extensive travel, and physical endurance, the 'rough and tumble of contests with contractors and labor unions' involved in the new construction and design of a plant would prohibit women's employment.[58]

Little's rhetorical position was broadly shared by male chemists working in the field. While stressing women's strength in all other lines of work connected with the chemistry lab, the chief chemist of the Calco Chemical Company voiced the general sentiment in 1919. 'It is impossible,' he said, 'to use women chemists on development work which has to be translated into plant practice by actual operation in the plant. This is the only limitation.' Other potential employers of women chemists elaborated on that particular theme by explaining that, 'research men must go into the Plant and manipulate all sorts of plant apparatus, direction [sic] foreign labor of every sort. You can readily see that a woman would be at a great disadvantage in this work . . .' Or that 'it often involves night work and almost always involves dealing with plant foremen and operators not easy to deal with.' The representative of the Grasselli Chemical Company's research department wrote in 1917 that chemical engineering involved 'large rough mechanical apparatus, . . . which work is usually carried on by unintelligent labor, in a good many cases the roughest kind of material.'[59] By establishing such notions as 'fact,' Little and other chemists succeeded in safely associating their work with both the male codes of the machine shop or the plant operation and managerial control. The power struggle in the workplace where matters of class were contested was a matter between men. Middle-class chemical engineers thus appropriated the male codes from the struggles of work place.

Despite the way women and men's engineering work was constructed, women did in fact probably do some of this work. Often discrepancies existed between job title and job content. Take the case of Glenola Behling Rose, whose title was chemist but who described her duties in 1920 as follows: 'I left the chemical dept. to go into the Dyestuffs Sales Dept. I have but one man over me and as his assistant, I am the Executive Office Supervisor of the Dyestuffs Technical Laboratory and have charge of all dealings with the chemical dept. such as deciding what dyes they shall go ahead to investigate & in what quantities, and keep track of their work in order to see whether they produce the dyes economically enough for us to market them. In a way I am the link between the research, the manufacturing and the selling of dyestuffs . . . As you will see a good deal of my work is supervisory.' With bachelors' degrees in geology and chemistry, and a Masters in chemistry, the highly qualified Glenola Rose felt she was technically well prepared for such job. In response to the question of what training she thought would be most beneficial to women entering her field, she replied that there was a 'need for a thorough foundation and a training with men,' by which she meant the task of managing men. And Florence Renick wrote that, in fact, 'I have had to deal considerably with labor of all kinds, mostly ignorant and many foreigners [sic] among them, and none of them but consider me "boss" so far as the laboratory is concerned.' On this particular point Jessie Elizabeth Minor, chief chemist at the Hamersley Manufacturing Company articulated women's ambivalence: 'There is still much masculine prejudice to combat. Many laboratories are not attractive looking. We come in contact with working men (which may be construed as an asset or liability).'[60] Thus, in these contexts, white women chemists actually did supervisory work and would have qualified as engineers according to the terms Little and other chemical engineering advocates had established for their engineering specialization. In all these instances, technical qualification or experience was less decisive in considerations for job assignments and promotions than the issue of supervision.

Women chemists were thus kept under job titles they actually had outgrown. Significantly, traffic between chemistry and chemical engineering also went in the other direction. Women who trained as chemical engineers ended up in lower-paid positions as chemists. Dorothy Hall (1894–1989) seemed to embody the success story of a woman advancing on the corporate ladder as a research and later chief chemist at GE. But with a Ph.D. in chemical engineering from the University of Michigan ('20) she was overqualified for her job. When asked, most employers said they thought women competent and excellent for research and analysis; few raised objections of a technical nature. But like Arthur D. Little, they all drew the line at work related to the plant operation. Thus in 1948 the Women's Bureau reported that most women who trained in chemical engineering were employed as chemists.[61]

In a limited way, chemical engineering and chemistry offered a niche to women students interested in engineering. Large numbers of women interested in the sciences in the U.S. and elsewhere flocked to the field of chemistry. The same held true for chemical engineering: more women graduated in chemical engineering than in any of the other engineering specializations. In fact, a higher proportion of female engineering students than of male engineering students majored in chemical engineering. But prominent chemical engineering advocates pushed for an explicitly male professional ethic by defining their discipline as an exclusive male domain that required supervisory skills. Tens of thousands of women chemists and chemical engineers were banished to chemical laboratories, where working conditions were dire and the pay was low.[62] In these contexts, the term chemist meant an ill-paid woman, while the title of chemical engineer often denoted a man in command of higher wages and managerial authority.

The male engineers' push for a high-status professional identity was in part a response to the enormous expansion of engineering work, which provided new opportunities to lower-class youths and sons of recent immigrants. The call for clear boundaries regarding class, however, resulted in a reinscription of male middle-class identity.

Census takers colluded in this through their unceasing efforts to find new categories for reliable enumeration; they sought to make classification consistent by excluding more and more groups of skilled workers from the category of engineer. Among those omitted were boat and steam shovel engineers, foremen of radio stations, engineers under thirty-five without a college education, and chemists. According to economists who have worked with the data, however, the statisticians made these adjustments without much attention to uniformity. These statistical and linguistic interventions did little to generate a satisfying set of data.[63] Nevertheless, historians have reproduced many of these figures, thereby accepting the categorizations provided without question. The definition of who would count as a true engineer and the production of statistics to justify this illusion worked together to create an illusory picture of a male, middle-class profession. The example of the chemical industry shows how linguistic construction and social practices made women invisible in engineering.

Organizing at Last

Women engineers responded to such strategies with stoicism but also with collective action. Before the First World War, the early generation of women engineers like Rickter, Wilson, Blatch, and Leiter had tried to gain access to the existing male organizations as individuals. They were rebuffed outright, granted secondary status, relegated to separate-but-unequal organizations, or segregated into different labor markets. A second generation of young women students and recent graduates including Lou Alta Melton, Hazel Quick, Elsie Eaves, Hilda Counts Edgecomb, and Alice Goff, who had found entry-level employment opportunities during the First World War, tried in 1919 to create a separate women's organization. But this failed. The post-suffrage generation championed the cause of women engineers with great enthusiasm. Yet none of the advocates of women's presence in engineering rallied to this feminist cause—even if they grounded their promotion of women engineers as professionals precisely in one of the important principles of modern feminism: as individuals women should be able to develop themselves to their fullest potential. All supported the notion that women had the freedom to choose whichever line of work suited their abilities, without the obligation to appeal to feminine propriety by arguing that such a choice was inspired by higher morals. But all ardently believed the engineering profession's promise of upward mobility. Resisting any direct association with the women's movement, therefore, they claimed instead that they just happened to have a knack for engineering as individuals. Success or failure was down to individual merit. This discourse was particularly prominent in 1943 when the government sought arguments to mobilize women for the war industries.[64] The majority of women engineers had internalized the values of corporate engineering, merit, and self-reliance.

The second generation saw their organizing efforts thwarted partly because they followed a logic of maintaining professional standards similar to that used by the male national organizations. Hence they excluded engineering students and working women engineers without formal education, such as non-collegiate draftsmen, chemists, and testing technicians, from their membership.[65] No doubt they had to do this in an effort to defend against sexism and to garner professional prestige. But emulating high professional standards prevented them from gathering the critical mass necessary for the success of such an organization.

In the same year that American women failed, British women succeeded. They established the Women's Engineering Society (WES), an inclusive organization that brought together women engineers with or without formal collegiate education as well as machinists who were skilled or semiskilled workers. The British successfully, albeit briefly, united across classes, in part because they did not adhere to the classic medical model of professionalism; instead they combined the tradition of upper-class business professionalism and trade associations with feminist ideals. In the end, [. . .] the British leaders also abandoned their policy of 'gender solidarity for male privilege and class advantage' and narrowed 'their focus to exclude the great mass of women who had entered the engineering trades during the First World War directly out of the working class.'[66] Between the world wars, when job opportunities virtually disappeared, American women engineers sought temporary shelter with their British colleagues through membership in the British WES and kept in touch through informal networks. Still clinging to the model of medicine, and having failed to create a critical mass of members, American women engineers of the interwar period turned instead to trying to shape public opinion. They did this partly through writing biographical sketches of each other according to well-established formulae, which stressed that with hard work and self-reliance women could—to paraphrase the title of Alice Goff's publication during these years—indeed become engineers. *[Goff's book is excerpted in Chapter 10. Ed.]* Thus American women engineers borrowed male models of meritocracy, but neither questioned the middle-class structure of American engineering nor campaigned for equal rights.

The final push towards organization in the United States did not come from the hundreds of thousands of women working in federal engineering jobs during the Second World War, from the informal networks of academically overqualified women engineers who had learned to be self-effacing during hard times, or from the women urban planners who had been nurtured in the Progressive era. It came once again—as it had in 1919—from young students and recent graduates eager to enter the job market and yearning for official recognition and respectability.

After the gap between rhetoric and reality had widened once again in the aftermath of the Second World War, American women engineers united at last in 1949. They did so long after other female professionals such as lawyers and doctors. An energetic and ambitious junior student leader on a scholarship, Phyllis Evans, won the support of the dean of women, Dorothy R. Young, and university counsel A. W. Grosvernor to organize the first meetings of women engineers at Drexel University in 1949. She and her colleagues organized over seventy young women engineering students from nineteen colleges on the East Coast. The goal was to make their 'voices . . . heard in the technological world' and to address inequities in engineering work. In the greater New York area, a group of women

Photograph of Phyllis Evans posing in overalls in Rosie-the-Riveter iconography illustrating a newspaper report on the first organizing efforts of women engineering graduates who had been disappointed by their employment opportunities after the peacetime conversion.
Reproduced from *The Christian Science Monitor*, 16 April 1949.

engineers who had been working in war-related industries—students from Cooper Union and City College of New York, coupled with graduates working nearby—also struck up conversations about their plight in the college libraries and Manhattan's coffee shops. Soon the long hidden tensions over leadership and the direction that professionalism should take for women engineers burst into public discussion.[67]

In the founding years, the student group at Drexel University in Philadelphia and a coalition of various groups in the greater New York area were in competition with each other for leadership. The origins of this contest concerned different professional strategies. Phyllis Evans—like so many other young women who had begun studying engineering during the war—was single, about to graduate, and facing unemployment. Echoing the governmental war rhetoric, she cherished high expectations for her future. Explaining her choice for engineering, she told a journalist about how her war experience as a cadet sergeant had inspired her to go into engineering. She wanted her future to be in military research. 'I want to build rockets and I want to go to Mars,' she said with youthful optimism. Establishment engineers such as Lillian G. Murad and Lillian Gilbreth, property owners who were steeped in the ethics of the patriarchal culture of family firms, opted for a more conservative approach, combining supervision and high professional standards. Although Lillian Gilbreth had been supportive of other women engineers, she did not favor a separate women's organization and was disinclined to head the SWE when it first sought a leader. She was most concerned about the bold effrontery of a separate organization and its feminist implications, and warned against blaming men for the difficulties women encountered in entering the field; instead she accused women of lacking credentials. 'The reason for women not being admitted into the National Engineering Society was not because they were women, but rather because they did not yet meet the qualification,' she said. But the elder Gilbreth was somewhat at odds with Dorothy Young, Drexel's dean of women students. Young both pushed for an activist strategy that confronted the inequity between men and women, but also struck a conciliatory note: women 'need to realize that it is necessary to work cooperatively with men in the larger field, planning together to abolish those inconsistencies that mar our democratic society,' she stated.[68]

The strategy of the young organization remained a balancing act: between the impatience of the students of Miller's generation, whose expectations had been raised by the government propaganda of Rosie-the-Riveter, and the cautious, conservative strategy of a previous generation of daughters and of wives, who were wedded to the patriarchal culture of family firms. The SWE tried to inspire younger women to go into engineering by taking individual women and their exemplary careers as models to be emulated. It did this through the establishment of medals and scholarship programs, the writing of biographical narratives, and the dissemination of pictures. In all of these, the organization stressed individual efforts. This was in direct contrast to the new corporate male ideal of team players that was promulgated by companies such as Dupont and General Motors. Thus the SWE highlighted merit and self-reliance, rather than becoming the collective movement for equality for which Nora Blatch had campaigned.

The SWE never resolved these conflicting goals. It sought to attract more young women to engineering schools, yet necessarily had to deal with the long-entrenched patriarchy of family firms, the exclusionary tactics of the professional organizations, and the discriminatory employment practices of corporations and the federal government. If it openly battled prejudice and sexism in engineering schools and practice, however, it risked frightening off prospective recruits.[69] Though the SWE was established in 1949 in the postwar period of 'adjustment,' when the government and corporations devised policies to push women back into the home, the organization would ride a new wave of ambiguous government encouragement during the Cold War. As in the former communist states like Soviet Union,

Group portrait of women attending the founding meeting of the American Society of Women Engineers at Green Engineering Camp of Cooper Union, N.J., May 27, 1950, as a response to the government's back-home campaigns just after the war.
SWE Archives, Walter P. Reuther Library, Wayne State University, used with permission.

the German Democratic Republic, and other Eastern European countries, the American military collaborated with corporations in actively recruiting women as technical personnel. 'Woman Power' campaigns were initiated after the Sputnik panic in the West in 1957. Despite its evocative image and name *Woman Power* (the title of a 1953 report issued by the National Manpower Council at Columbia University) had little to do with feminist calls for equal rights: it shied away from controversial issues such as equal pay for equal work and job discrimination. It neither upset gender hierarchies nor helped foster a separate women's culture. Instead, it provided the conservative part of the women's movement a certain legitimacy, since it stressed that women, if they worked hard, could alleviate national labor shortages.[70] This new society successfully gathered the critical mass necessary for such a separate organization despite its high professional standards because the cold war had created such a high demand for technical personnel where numbers had been lacking three decades earlier. The British women engineers had been so successful at organizing their sisters because they had been building bridges with both the women's movement and the rank-and-file of their profession. The truth was that in the United States women engineers' individual opportunities were part of America's military-industrial complex and highly depended on it. At times its doors might be open to women, but it almost always reproduced old patriarchal patterns in a new corporate context.

'Woman Power' and Daughters of Martha

The government's campaign of Woman Power was a borrowed model. So was Lillian Gilbreth's appropriation of Rudyard Kipling's 1907 poem 'The Sons of Martha.'[71] At the 1961 opening of the Society of Women Engineers' new headquarters in the United Engineering Building, Gilbreth made the poem relevant to the needs of women engineers by entitling her speech 'The Daughters of Martha.' The modernist United Engineering Building towered high in New York and expressed the coming of age of the engineering professions, but the new headquarters Gilbreth was about to open were tiny and symbolic of women's place in the profession. By invoking Kipling's allegory, Gilbreth sought to empower women in the technical professions through the values of service, sacrifice, and self-reliance. In so doing, she showed the marginalized place of women in a male technical world and the hardship of women who labored on the lower rungs of the profession as rank-and-file engineers and corporate workers. She thus put forward a model that doubled the burden on women who aspired to be engineers: they were expected to

Photograph taken as part of the campaign to recruit women for engineering positions for the U.S. Military as the Cold War heats up in 1953. In this photograph, gender relations are preserved rather than subverted by the photographer's frog-eyed frame stressing masculine features of the woman in charge and the subordinate position of her reclining colleague.
Courtesy of the Schlesinger Library, Radcliffe Institute, Harvard University, used with permission.

make sacrifices both by virtue of their sex and because of their profession.[72] By the 1960s, Gilbreth's professional model of 'simple service simply given,' or, as she had advised earlier, 'helping others express themselves [as] the truest self expression,' was out of date for women.[73] Her call for the inclusion of women in the profession was based on her own long career, and on the trying experiences of many other women engineers. But principally it rested on a conservative notion that the best way forward for women professionals generally was to be ever stoical and always overqualified. Despite her attempt to redefine the place of women at the center of engineering, her celebration of service, sacrifice, and self-reliance reinforced some very traditional notions about women's proper place in engineering. Her employment of the Kipling's poem threatened to become a failed allegory. When male engineers used the poem, it could be mobilized to appropriate working-class badges of manliness or to symbolize them as underdogs, but when women mobilized the poem for their cause the figure of Martha turned into an image of subordination, stoicism, and lack of advancement. Men perhaps could pass 'down' but women could rarely pass 'up' the cultural hierarchies as women found here and elsewhere.

Women engineers in the military-industrial complex or in the patriarchal culture of family firms had no appealing role models except the problematic image that Lillian Gilbreth supplied. Outside the military-industrial complex and the patriarchal family firms, however, women were building their own structures. The women of the Progressive era who participated in the women's reform movement also helped to shape an alternative women's technical culture that was nurtured by women's traditions.[74] Throughout the country, from Boston to San Francisco, women reformers helped build the public infrastructure of the civic improvements movement as private citizens rather than as corporate employees. Highly organized in private philanthropic organizations such as the General Federation of Women's Clubs, these women reformers campaigned for what they called municipal housekeeping. They conducted surveys, drew up plans for urban infrastructures, pushed for better housing, and helped finance public facilities from streetlights to sewer systems. They forged coalitions with local politicians, architects, civic leaders, and professional women such as Ellen Swallow Richards *[Richards is profiled in Chapter 5. Ed.]*, Alice Hamilton, and Ruth Carson in public health, science, and social research respectively. The women of the Progressive movement became urban planners of the modern age.[75] There existed, therefore, a rich female heritage of building. But this was not available to women engineers working in family firms or military and corporate industries. In fact, the women of these separate technical cultures never met or sought to bridge the gap between them.

Notes

1. An earlier version of this chapter appeared in Ruth Oldenziel, *Making Technology Masculine: Men, Women and Modern Machines in America 1870–1945*, Amsterdam University Press, 1999. The author is most grateful to Karin Zachmann for sustained critical readings of earlier drafts.
2. Samuel Florman, "Engineering and the Female Mind," *Harper's* (February 1978): 57–64.
3. Florman, "Engineering and the Female Mind," 61.

4. Peter F. Meiksins, "Engineers in the United States: A House Divided," in *Engineering Labour. Technical Workers in a Comparative Perspective*, edited by Peter Meiksins and Smith Chris (London: Verso, 1996), 61–97 and his "Professionalism and Conflict: The Case of the American Association of Engineers," *Journal of Social History* 19, no. 3 (Spring 1986): 403–22.

5. Bruce Seeley, "Research, Engineering, and Science in American Engineering Colleges, 1900–1960," *Technology and Culture* 34 (1993): 344–86; Peter Lundgreen, "Engineering Education in Europe and the U.S.A., 1750–1930: The Rise to Dominance of School Culture and the Engineering Professions," *Annals of Science* 47 (1990): 33–75.

6. Unless otherwise indicated, this article is based on my survey of all American accredited engineering schools prior to 1945. A query was sent to 274 schools. This correspondence was followed up by letters to registrar's offices, engineering colleges, university archives, and alumni associations. In total, 175 schools answered: 82 reported on women graduates. Of the 99 schools that did not respond, only 10 schools appear to have awarded a significant number of degrees to women. Based on this survey and other sources, over 600 women in engineering have been identified by name. *American Women Engineering Graduates Papers*, Author's Personal Collection (AWEG Papers, hereafter).

7. Examples of hostility: Vera Jones Mackay in *100 Years: A Story of the First Century of Vanderbilt University School of Engineering, 1875–1975* (Nashville: Vanderbilt Engineering Alumni Association, 1975), 116–9; Marion Monet (MIT '43) personal interview March 23, 1990; Author's telephone interview with Dorothy Quiggle November 14, 1989, and *MIT Survey*, example of aloofness: Eleonore D. Allen (Swarthmore '36), "Lady Auto Engineer Her Ideas Irreparable," *New York World Telegram & Sun* (August 14, 1961). On post-suffrage generation of women professionals, see: Nancy Cott, *The Grounding of Feminism* (New Haven: Yale University Press, 1987), Chapter I and Introduction.

8. Rosalyn Rosenberg, *Beyond Separate Spheres. Intellectual Roots of Modern Feminism* (New Haven: Yale University, 1983); Louise Michele Newman, ed., *Men's Ideas/Women's Realities* (New York: Pergamon Press, 1985); John M. Staudenmaier, S. J., *Technology's Storytellers. Reweaving the Human Fabric* (Cambridge, MA: MIT, 1985); Margaret W. Rossiter, *Women Scientists in America: Struggles and Strategies to 1940* (Baltimore: The Hopkins University Press, 1982).

9. Author's interview with Nora Blatch's daughter, Rhoda Barney Jenkins, September 1989; Ellen DuBois, ed., "Spanning Two Centuries: The Autobiography of Nora Stanton Barney," *History Workshop Journal* 22 (1986): 131–52, p. 134. Cf. Suzy Fisher, "Nora Stanton Barney, First U.S. Woman CE, Dies at 87," *Civil Engineering* 41 (April 1971): 87.

10. Lillian Gilbreth, "Marriage, a Career and the Curriculum," typewritten manuscript (probably 1930s), Lillian Gilbreth Collection, NHZ 0830-27, Box 135, Department of Special Collections and Archives, Purdue University Library, Lafayette, IN (Gilbreth Papers hereafter). See also "American Women Survey

Their Emancipation. Careers Are Found an Aid To Successful Married Life," *Washington Post* (August 16, 1934); Edna Yost, *Frank and Lillian Gilbreth, Partners for Life* (New Brunswick: Rutgers University Press, 1949) and Yost's description in *American Women of Science* (New York: Frederick A. Stokes, 1943); Ruth Schwartz Cowan, *Dictionary of Notable Women. Modern Period* (Cambridge: Belknap Press, 1980), s.v., "Gilbreth." The notion of "borrowed identity" comes from Margot Fuchs, "Like Fathers-Like Daughters: Professionalization Strategies of Women Students and Engineers in Germany, 1890s to 1940s," *History and Technology* 14, 1 (1997): 49–64.

11. See Rossiter's excellent *Women Scientists in America* (1982), 248. The sequel *Women Scientists in America, Before Affirmative Action, 1940–1972* (Baltimore: The Johns Hopkins University Press, 1995) is equally pathbreaking and continues to be an inspiration.

12. *First Century of Vanderbilt University School of Engineering, 1875–1975*, 116–9; Correspondence Vanderbilt University, Special Collections University Archives, Nashville, TN, AWEG Papers.

13. Ovid Eshbach, Technological Institute, Northwestern University, Chicago, IL, February 15, 1945 with Woman's Bureau, RG 86, Box 701, Woman's Bureau, National Archives, Washington, D.C., (WB NA hereafter). Cf. Juliane Mikoletzky, "An Unintended Consequence: Women's Entry into Engineering Education in Austria," *History and Technology* 14, 1 (January 1997): 31–48. Fuchs, "Like Fathers-Like Daughters."

14. Surveying many unusual fields of women's employment, Miriam Simons Leuck reported on a great many widows, some of whom were engineers: "Women in Odd and Unusual Fields of Work," *AAAPPS* 143 (1929): 166–79; U.S. Bureau of the Census, *Census 1890* (Washington, D.C.: Government Printing Office, 1890).

15. A., "Women Engineers," Letter to the Editor, *Professional Engineer* (May 1922): 20; "Roebling Memorabilia" *The New York Times* (October 4, 1983); Gustave Lindenthal, Letter to the Editor, "A Woman's Share in the Brooklyn Bridge," *Civil Engineering* 3, 3 (1933): 473; Alva T. Matthews, "Emily W. Roebling, One of the Builders of the Bridge," in *Bridge to the Future: A Centennial Celebration of the Brooklyn Bridge*, edited by Margaret Larimer, Brooke Hindle, and Melvin Kramberg *Annals of the New York Academy of Sciences* (1984), 63–70; for a biography on Roebling, see Marilyn Weigold's study, *The Silent Builder: Emily Warren Roebling and the Brooklyn Bridge* (Port Washington, NY: Associated Faculty Press, 1984). For an appreciation of her work by the engineering community, see "Engineers Pay Tribute to the Woman Who Helped Build the Brooklyn Bridge," (address by David B. Steinman for the Brooklyn Engineers Club, May 24, 1953), reprinted in *The Transit of Chi Epsilon* (Spring–Fall 1954): 1–7.

16. For a treatment of Gilbreth's work, see Martha Moore Trescott "Lillian Moller Gilbreth and the Founding of Modern Industrial Engineering," in *Machina Ex Dea: Feminist Perspectives on Technology*, ed. by Joan Rothschild (New York: Pergamon Press. 1983), 23–37. "Marriage, a Career and the Curriculum"; "American Women Survey Their Emancipation"; Yost *Frank and Lillian*

Gilbreth and *American Women of Science*. Yost was one of the Gilbreths' most important promoters and popularizers. Cowan in *Notable American Women*, s.v., "Gilbreth."

17. Correspondence Special Collections Department, University Archives, University of Nevada, Reno; correspondence with New Jersey Institute of Technology, Alumni Association, Newark, NJ, (AWEG Papers); Carolyn Cummings Perrucci, "Engineering and the Class Structure," in *The Engineers and The Social System*, eds. Robert Perrucci and Joel Gerstl (New York: John Wiley and Sons, 1969), 284, Table 3.

18. Correspondence Cornell University Library, Department of Manuscripts and University Archives (AWEG Papers); Eve Chappell, "Kate Gleason's Careers" *Woman's Citizen* (January 1926): 19–20, 37–8; *DAB*, s.v. "Gleason"; "A Woman Who Was First," *The Cornell Alumni News* (January 19, 1933): 179 with excerpts from *The Cleveland Plain Dealer* and *The New York Tribune*; Leuck, "Women in Odd and Unusual Fields of Work," 175; *The Gleason Works, 1865–1950* (n.p., 1950); ASME. *Transactions* 56, RI 19 (1934), s.v., "Gleason." Cf. Christopher Lindley in *Notable American Women. Supplement* 1 (1934), s.v., "Gleason."

19. Judith S. Mcllwee and Gregg J. Robinson, *Women in Engineering: Gender, Power and Workplace Culture* (Albany: State University of New York Press, 1992).

20. DuBois, "Spanning Two Centuries," 150.

21. Terry Kay Rockefeller in *Notable American Women* (1980), s.v., "Barney."

22. "Barnard Girls Test Wireless Phones," *The New York Times* (February 23, 1909):7:3.

23. In *Inventing American Broadcasting, 1899–1922* (Baltimore: The Johns Hopkins University, 1987), 167–7, Susan J. Douglas writes that Nora Blatch's "contributions to early development of voice transmission have been either completely ignored or dismissed", (p. 174).

24. "Warns Wives of 'Careers'," *The New York Times* (July 28, 1911) 18:3 and response by Ethel C. Avery, "Suffrage Leaders and Divorce" (July 31, 1911) 6:5. See also Margaret W. Raven, a graduate from MIT ('39) in General Science and Meteorology, Association of MIT Alumnae, Membership Survey, 1972, MIT, Institute Archives, Cambridge, MA; and Mcllwee and Robinson, *Women in Engineering*.

25. David Noble, *A World Without Women: The Christian Clerical Culture of Western Science* (New York: Alfred A. Knopf, 1992).

26. Edna May Turner, "Education of Women for Engineering in the United States, 1885–1952," (Ph.D. diss., New York University, 1954). This valuable and pioneering dissertation contains little analysis or biographical information beyond the statistics.

27. Correspondence with University Archives, Ohio State University, Ames *[sic]*, OH; Special Collections and Center for Southern History and Culture, University of Alabama, Tuscaloosa, AL; correspondence with Stanford Alumni Association, Stanford, CA (AWEG Papers). Robin found a similar trend for the post-World War era. "The Female in Engineering," in *The Engineers and the Social*

System, 203–218. Cf. "Report of the Committee on Statistics of Engineering Education," *Proceedings of the Society for the Promotion of Engineering Education* 10 (1902): 230–57, p. 238, Table I.

28. Tom S. Gillis, roommate of Henry M. Rollins, Hutson's son, letter to author, October 21, 1989; correspondence Texas A&M University College Station, TX; University of Missouri-Rolla, Library and Learning Resources, University of Missouri-Rolla, Rolla, MO, (AWEG Papers); Lawrence O. Christensen and Jack B. Ridley, *UM-Rolla: A History of MSM/UMR* (Missouri: University of Missouri Printing Services, n.d.), 93–4; and his "Being Special: Women Students at the Missouri School of Mines and Metallurgy," *Missouri Historical Review* 83, 1 (October, 1988): 17–35; Frances A. Groff, "A Mistress of Mechanigraphics," *Sunset Magazine* (October 1911): 415–8.

29. Clipping file. Alumni Office, Swarthmore College, Swarthmore, PA; for further information: *Yearbook* (1942) (AWEG Papers); Edward M. Tuft, "Women in Electronics," *National Business Women* (November 1956). See also: Olive W. Dennis to Marguerite Zapoleon, September 3, 1947, Women's Bureau, Bulletins, RG no. 223-225, WB, NA.

30. Sally Hacker, "The Mathematization of Engineering: Limits on Women and the Field," in *Machina Ex Dea*, 38–58. After the Second World War, competence in mathematics was the single most common denominator in women's motivation to go into engineering. Martha Moore Trescott. "Women Engineers in History: Profiles in Holism and Persistence," in *Women in Scientific and Engineering Professions*, eds. Violet B. Haas and Carolyn C. Perrucci (Ann Arbor: The University of Michigan Press, 1984), 181–205. More work needs to be done on women's education in mathematics at the secondary school level, but see Warren Colburn, "Teaching of Arithmetic. Address before the American Institute of Instruction in Boston, August 1830," in *Readings in the History of Mathematics Education* (Washington, D.C.: National Council of Teachers in Mathematics, 1970), 24–37; Ruth Oldenziel, "The Classmates of Lizzie Borden" (University of Massachusetts: unpublished manuscript, 1982). See also Robert Fox and Anna Guagnini, "Classical Values and Useful Knowledge: The Problem of Access to Technical Careers in Modern Europe," *Daedalus* 116, No. 4 (1987): 153–71. For an excellent discussion on the issue during the 1970s and 1980s, McIlwee and Robinson, *Women in Engineering*.

31. Nora Blatch Photo Album, Courtesy of her daughter Rhoda Barney Jenkins, Greenwich, CT (Barney-Jenkins Papers, hereafter); Records of the Women's Bureau, Women's Bureau Bulletin 223–225; RG, Box 701, WB NA; "Rice Women Engineering," *Rice Engineer* (January 1986): 18–23. *Women Engineers and Architects* (March 1938): 1, Box 131 folder "Women Engineers and Architects, 1938–1940," Society of Women Engineers Collection, Walter Reuther Library, Wayne State University, Detroit, MI (SWE Collection, hereafter); correspondence Woodson Research Center, Rice University, Houston, TX.

32. Karl Drews, "Women Engineers: The Obstacles in Their Way," *Scientific American, Supplement* 65 (March 7, 1908): 147–8.

33. Barbara Drygulski Wright in her introduction to *Women, Work, and Technology Transformations* (Ann Arbor: University of Michigan Press, 1987), 16–7.

34. John W. Upp, "The American Woman Worker," *The Woman Engineer 1* and "American Women Engineers," *The Woman Engineer 1*, 11 (June 1922): 156; 186–88.

35. *Report of the Registrar of the University of Michigan, 1926–1941*; correspondence Office of Development, Bentley Historical Library, University of Michigan, Ann Arbor, MI (AWEG Papers).

36. College of Engineering and Applied Science; Special Collections, University of Colorado, Boulder, CO; correspondence University of Nevada, Special Collections Department, Reno, NV, Ohio Northern University, Alumni Office, Ada, OH; *Alumni Directory* (1875–1953), University of Minnesota; *The Minnesota Techno-Log* (May 1925): 11; correspondence University of Minnesota, University Archives, Minneapolis, MN (AWEG Papers).

37. Correspondence with Alumni Records Office, College of Engineering, Bentley Historical Library, University of Michigan, Ann Arbor, MI; correspondence with Bertha L. Ihnat; Ohio State University, *Commencement Programs* (1878–1907), Ohio State University, Ames *[sic]*, OH (AWEG Papers) clipping file. General Electric Company, GE Hall of History Collection; Goff, *Women Can Be Engineers*; Elsie Eaves, "Wanted: Women Engineers," *Independent Woman* (May 1942): 132–3, 158–9, p. 158.

38. Correspondence with Alumni Association, Iowa State University, Ames, IA (AWEG Papers); Adelaide Handy, "Calculates Power Transmission for General Electric Company," *The New York Times* (October 27, 1940).

39. Dennis to Marguerite Zapoleon, September 3, 1947, RG, 223–225, Box 701, WB NA.

40. Nora Stanton Barney letters to the editor, "Industrial Equality for Women," *N.Y. Herald-Tribune* (April 21, 1933) and "Wages and Sex," *N.Y. Herald-Tribune* (July 21, 1933).

41. Elsie Eaves to Mary Esther Poorman, November 8, 1933; Elsie Eaves to Virginia A. Swary, March 2, 1936; Elsie Eaves to Jane Hall, November 10, 1938, Box 146, folder "Earliest Efforts to Organize, 1929–1940," SWE Collection.

42. Box 146, folder "Earliest Efforts to Organize, 1920–1940," SWE Collection.

43. Juliet K. Coyle, "Evolution of a Species—Engineering Aide," *US Woman Engineer* (April 1984): 23–4; Robert McMath, Jr. et al., *Engineering the New South. Georgia Tech, 1885–1985* (Athens: The University of Georgia Press, 1985), 212; correspondence John D. Akerman with Curtiss-Wright Corporation, 1942–43, University Archives, University of Minnesota, Minneapolis, MN; Curtiss-Wright Engineering Cadettes Program Papers, Archives of Women in Science and Engineering, Iowa State University, Ames, IA; correspondence Georgia Institute of Technology, Archives and Records Department, Atlanta, GA; "Engineering Aide, Curtis Wright Program Follow-up of Cadette trained at Rennselaer Polytechnic," March 22, 1945, Rennselaer Polytechnic Library, NY, (AWEG Papers).

44. C. Wilson Cole, "Training of Women in Engineering," *Journal of Engineering Education* 43 (October 1943): 167–84; E. D. Howe, "Training Women for Engineering Tasks," *Mechanical Engineering* 65 (October 1943): 742–4; R. H. Baker and Mary L. Reimold, "What Can Be Done to Train Women for Jobs in Engineering," *Mechanical Engineering* 64 (December 1942): 853–5; D. J. Bolanovich, "Selection of Female Engineering Trainees," *Journal of Educational Psychology.* (1943) 545–53; "Free-Tuition in Courses Engineering for Women," *Science* 95, 2455, Suppl. 10 (January 9, 1942): 10; "Training for Women in Aeronautical Engineering at the University of Cincinnati," *Science* 97 (June 18, 1943): 548–9.

45. *Training in Business and Technical Careers for Women* (Ohio: The University of Cincinnati, 1944); Harriette Burr, "Guidance for Girls in Mathematics," *The Mathematics Teacher* 36 (May 1943): 203–11. Margaret Barnard Pickel, "A Warning to the Career Women," *The New York Times Magazine* (July 16, 1944): 19, 32–3 and Malvina Lindsay, "The Gentler Sex. Young Women in a Hurry," *Washington Post* (July 20, 1944).

46. Olive W. Dennis to Marguerite Zapoleon, September 3, 1947, WB, NA.

47. Many examples may be found in the Records of the Women's Bureau Bulletins, RG 86, WB NA.

48. Coyle, "Evolution of a Species"; US Department of Labor, Woman's Bureau, *The Outlook for Women in Architecture and Engineering*, Bulletin 223 no. 5 (Washington, DC: Government Printing Office, 1948); U.S. Department of Labor, Woman's Bureau, "Employment and Characteristics of Women Engineers," *Monthly Labor Review* (May 1956): 551–6.

49. Excerpts from Olive W. Dennis, "So–Your Daughter Wants to be a Civil Engineer," Box 701, WB NA; excerpts from letter, *Baltimore and Ohio Magazine* (September 1940): 30.

50. Olive W. Dennis, Clipping File, Baltimore and Ohio Railroad Museum, Baltimore, MD.

51. "Women AIME Members Contribute Their Share in Engineering War," *Mining and Metallurgy* 23 (November 1942): 580–1.

52. For gender differences in the professions: Joan Brumberg, and Nancy Tomes, "Women in the Professions: A Research Agenda for American Historians," *Reviews of American History* 10, 2 (June 1982): 275–96; Barbara F. Reskin and Polly A. Phipps, "Women in Male-Dominated Professional and Managerial Occupations," in *Women Working: Theories and Facts in Perspective*, eds. Ann H. Stromberg and Shirley Harkness (Mountain View, CA: Mayfield, 1988); Barbara F. Reskin and Partricia A. Roos, *Job Queues, Gender Queues: Explaining Women's Inroads into Male Occupations* (Philadelphia: Temple University Press, 1990). See also, Andrew Abbott, *The Systems of Profession: An Essay on the Division of Expert Labor* (Chicago: Chicago University Press, 1988), 98–111.

53. The number of women in the professional organizations was reported in Woman's Bureau, *The Outlook for Women,* 22. In 1946, women accounted for six out

of every thousand members in the ASCE and AIEE. Nine participated in the AIChE, 16 in the ASME, and 21 in the AIME. These figures roughly resemble the data gathered in the survey. New research should focus on local engineering organizations, however.

54. A. Michal McMahon, *The Making of a Profession: A Century of Electrical Engineering in America* (New York: The Institute of Electrical and Electronics Engineers Press, 1984), 58–9.

55. Anson Marston to Hilda Counts, May 6, 1919, Society of Women Engineers Papers, Box 146, folder "Earliest Efforts to Organize, 1918–1920," SWE Collection. On her suit and efforts to rally support for her case, see W. W. Pearse to Nora S. Blatch, January 20, 1915 and another from Ernest W. Schroder to Nora S. Blatch, January 20, 1915, Barney-Jenkins Papers. "Mrs. De Forest Loses Suit," *The New York Times* (January 22, 1916): 13; Reports on the case appeared also on Saturday January 1, 1916: 18, "Mrs. De Forest Files Suit"; January 12, 1916: 7; "Old Men Bar Miss Blatch"; and in *The New York Sun* January 1, 1916: 7, "Mrs. De Forest, Suing, Tells her Real Age." Blatch's employment history may be found in *Notable American Women* (1980), s.v., "Barney"; DuBois, "Spanning Two Centuries," 148; Speech "Petticoats and Slide Rules", 6, Box 187, folder "Miscellaneous Correspondence, Elsie Eaves", SWE Collection. The speech was published under the same title in a slightly altered form in *The Midwest Engineer* (1952).

56. "Women Engineers–Yesterday and Today," *The Bent of Tau Beta Pi* (Summer 1971): 10–2; correspondence University Library, University Archives, University of Illinois at Urbana-Champaign, Urbana, IL, (AWEG Papers).

57. Rossiter, *Women Scientists in America* (1982), 77 and Chapter 4. For a sample of the literature on women in the professions see Barbara Melosh, *The Physicians' Hand: Work Culture and Conflict in American Nursing* (Philadelphia: Temple University Press, 1982); Brumberg and Tomes, "Women in the Professions."

58. *Dictionary of Occupational Titles* (1939); Arthur D. Little to Beatrice Doerschuk, February 6, 1922, Bureau of Vocational Information, Schlesinger Library, Radcliffe College, Cambridge, MA, microfilm [BVI, hereafter], reel 12. His opinion was extensively quoted in a section on opportunities in chemical engineering in the Bureau of Vocational Information, *Women in Chemistry: A Study of Professional Opportunities* (New York: Bureau of Vocational Information, 1922), 60–1. Terry S. Reynolds, "Defining Professional Boundaries: Chemical Engineering in the Early 20th Century," *Technology and Culture* 27 (1986): 694–716, p. 709. On Little, see also David F. Noble, *America By Design* (New York: Oxford University Press, 1977), 124–5.

59. Calco Chemical Company, M. L. Crossley to Emma P. Hirth, December 24, 1919; National Aniline & Chemical Company, C. G. Denck to Emma P. Hirth, December 22, 1917; A. P. Tanberg, Dupont to Emma P. Hirth, August 30, 1921; L. C. Drefahl Grasselli Chemical Company to Emma P. Hirth, December 20, 1917; all BVI reels 12 and 13.

60. Mrs. Glenola Behling Rose, chemist at Dupont Company, questionnaire, February 1920, BVI reel 12; Florence Renick, questionnaire, February 1920, BVI reel 13; Jessie Elizabeth Minor, questionnaire, January 12, BVI reel 12.

61. US Department of Labor, Women's Bureau, *The Occupational Progress of Women, 1910 to 1930*, Bulletin 104 (Washington, D.C.: Government Printing Office, 1933); *The Outlook for Women*, 46.

62. US Department of Interior, Office of Education, *Land-Grant Colleges and Universities* (Washington, DC: Printing Office, 1930), 805; Ruth Oldenziel, "Gender and the Meanings of Technology: Engineering in the US, 1880–1945," (Ph.D. diss., Yale University, 1992), Fig. 6.

63. David M. Blank and George J. Stigler, *The Demand and Supply of Scientific Personnel*, General Series, 62 (New York: National Bureau of Economic Research, 1957) 4, 8–9, 10-2, 87, 192.

64. Cott, *The Grounding of Feminism*, Chapter 1 and Introduction.

65. Box 146, folder 'Earliest Efforts to Organize, 1918–1920,' SWE Collection.

66. Carroll Pursell, "Am I a Lady or an Engineer? The Origins of the Women's Engineering Society in Britain, 1918–1940", in Annie Canel, Ruth Oldenziel, and Karin Zachmann, eds. *Crossing Boundaries, Building Bridges: Comparing the History of Women Engineers 1870s–1990s*. (Amsterdam: Harwood Academic Publishers, 2000). See also Crystal Eastman, "Caroline Haslett and the Women Engineers," *Equal Rights* 11/12 (10 April 1929): 69–70.

67. "Origins of the Society by Phyllis Evans Miller," Box 147, SWE Collection; "Girls Studying Engineering See Future for Women in These Fields," *The Christian Science Monitor* (Saturday, April 16, 1949): 4; "New Members of the Women's Engineering Society" *Women's Engineering Society* (1950), 315.

68. "Girls Studying Engineering See Future"; Marta Navia Kindya with Cynthia Knox Lang, *Four Decades of the Society of Women Engineers* (Society of Women Engineers, n.d.), 12; Lillian G. Murad to Katharine (Stinson), June 8, 1952, Box 146, folder "SWE history 1951–1957," SWE Collection.

69. See also Carroll W. Pursell, *The Machine in America: A Social History of Technology* (Baltimore: The Johns Hopkins University Press, 1995), 310, for a passing, but insightful, remark on this issue.

70. Rossiter, *Women Scientists in America* (1995), 28, 59, and Chapter 2.

71. Lillian M. Gilbreth, "The Daughters of Martha," speech before the Society of Women Engineers at the opening of the headquarters in the United Engineering Building, SWE celebration banquet, 1961. Box 24, Gilbreth Papers.

72. Today the SWE's office is overcrowded and understaffed, occupying a tiny space in an otherwise imposing building where all engineering societies reside together near the United Nations Headquarters in New York City. *[The Society of Women Engineers moved out of the United Engineering Center in 1993 to larger quarters on Wall Street in lower Manhattan and later moved to Chicago. The United Engineering Center property was subsequently demolished for a new development. Ed.]*

73. "Marriage, a Career and the Curriculum," Box 135, Gilbreth Papers.
74. For a provocative inquiry, see: Pamela Mack "What Difference has Feminism Made to Engineering in the 20th Century?" in *Feminism in 20th Century Science, Technology, and Medicine* eds. Angela N.H. Creager, Elizabeth Lunbeck, and Londa Schiebinger (Chicago: University of Chicago Press, 2001).
75. Mary Ritter Beard, *Woman's Work in Municipalities* (New York: Arno Press, 1972 [1915]).

II

Pioneers and Trailblazers

This section, with one exception, focuses on the experiences and contributions of individual women engineers. It opens with Margaret Ingels' entertaining 1952 lecture "Petticoats and Slide Rules," describing a parade of women entering the engineering profession from 1886 through the early 1920s, and continues with more extensive profiles of pioneering women engineers of the late nineteenth and early twentieth centuries, culminating with a group portrait of women engineering society presidents at the beginning of the twenty-first century. The exception is Jennifer Light's analysis of the contribution of women to the development of the first electronic computer, the ENIAC, in Chapter 11. The chapters in this section draw from a variety of sources, including women's magazines, professional society publications, a college alumni magazine, and scholarly journals dating from the 1920s to the early twenty-first century. They are presented, more or less, in chronological order of the lives of the women profiled.

Chapter 4 serves as an introduction to this section. In preparation for a speech to the Western Society of Engineers in Chicago in 1952, Margaret Ingels contacted engineering colleges, professional societies, and the U.S. Department of Labor in an attempt to identify early women engineers. She describes a parade of "petticoat engineers" and their accomplishments, and urges women engineers that follow to "widen the trails blazed for her." Ingels, herself a pioneering engineer and the first woman to graduate from the University of Kentucky's College of Engineering, is profiled in Chapter 10.

Martha Trescott's Chapter 5 was first published in 1990 in *Women of Science: Righting the Record*, and emphasizes the holistic approach brought to various fields of engineering by pioneering women. Trescott collected historical information and surveyed and interviewed many women engineers in the 1970s for a book project. Her profiles of sanitary chemist Ellen Swallow Richards and industrial psychologist Lillian Gilbreth, neither of whom trained as engineers, are excerpted here, and focus on the contributions of these two women to the development of entirely new fields of engineering study and practice that were nevertheless overshadowed by their husbands and proteges. Trescott's celebration of Gilbreth's accomplishments contrasts with Oldenziel's more critical view of Gilbreth in Chapter 3. While Trescott attributes Gilbreth's lack of recognition to historical circumstance, Oldenziel faults Gilbreth for taking a self-effacing approach instead of asserting herself more strongly on behalf of women.

In Chapter 6, Alva Matthews, another engineer author, describes the role of Emily Roebling, wife of Washington Roebling, chief engineer of the Brooklyn Bridge, in the construction of that New York landmark. After her husband was incapacitated early in the project by "caisson disease," what we today know as "the bends," or decompression sickness, Emily became his go-between, meeting with his assistants, overseeing progress, and directing construction. We can only speculate on the true extent of her involvement in the project. Matthews' piece was written for the New York Academy of Sciences' celebration of the centennial of the bridge's construction in 1984.

The colorful life of Kate Gleason, the first, and so far only, woman to have an engineering school in the U.S. named after her (at the Rochester Institute of Technology), is described in Chapter 7. Gleason attended Cornell to study mechanical engineering, but was called back home before completion of a degree to help her father in the family machine tool business. She kept the books and traveled the United States and Europe in the 1890s to sell the firm's equipment, eventually becoming the first woman member of the American Society of Mechanical Engineers. After a successful career at the Gleason Works, she went on to become a pioneering bank president and real-estate developer. Eve Chappell's interview with Gleason for *The Woman Citizen* dates from 1926.

Another nineteenth-century engineer, Bertha Lamme, is profiled in Chapter 8. An 1893 graduate of Ohio State University with a degree in mechanical engineering and the first woman engineer employed by Westinghouse Electric Works, Lamme contributed to the development of the electrical generation and transmission grid in the U.S., but like many of her contemporaries her brief career ended when she married another engineer at the firm. Jeff Meade's profile was first published in the American Society for Engineering Education's *Prism* magazine in 1993.

James Brittain, a historian of the electrical industry in America, first published Chapter 9 in the *IEEE Transactions on Education* in 1985. He documents the career of Edith Clarke from schoolteacher to human computor to electrical engineer to engineering professor in the first half of the twentieth century. Clarke's long and circuitous route to an engineering career illustrates the many challenges faced by women at that time. With a math degree, she tried teaching, but decided to go into engineering, eventually becoming the first woman to complete an electrical engineering degree at MIT. After joining General Electric (GE), Clarke developed new techniques to solve problems related to long-distance electrical power transmission, and went on to become a professor at the University of Texas for another 11 years after retirement from GE in 1945.

Alice Goff's 1946 book *Women Can Be Engineers* celebrates the accomplishments of a dozen women engineers and six women in closely related fields. A civil engineer herself, Goff was engaged in early efforts to attract more women to the profession. Goff's entertaining profiles humanize their subjects by commenting on their childhood, hobbies and family life as well as their professional accomplishments. Included in Chapter 10 are her foreword and biographies of Olive Dennis, a draftsman and engineer on the Baltimore and Ohio Railroad, Margaret Ingels (author of Chapter 4), who worked for the Carrier Corporation in the early days of air conditioning, and Elsie Eaves, another woman with a non-traditional engineering career employed by McGraw-Hill, publisher of the *Engineering News Record*. Ingels became the first female graduate of the engineering school at the University of Kentucky in 1916; Dennis and Eaves both received engineering degrees in 1920. Dennis designed many improvements in rail car design to increase passenger comfort in her role as "engineer of service" for the B&O. During her long career in air conditioning, Ingels worked in research as well as advertising, sales, and public relations, and was responsible for developing the "effective temperature" scale incorporating humidity and air movement into an equation to predict human comfort. Eaves spent 37 years at McGraw-Hill where she pioneered the development of construction cost indexes and compiled national data on construction projects and needs as manager of the business news department.

Chapter 11 departs from profiling individual women to examine women's contributions to the development of the first electronic computer, the ENIAC, during World War II. Historian Jennifer Light describes how the job of computer programmer originated as feminized clerical work, evolving from the task of performing mathematical calculations, and how women have subsequently been omitted from the standard histories of computing. She also highlights the contradictions between purported opportunities for women during the war years and the limited options actually open to them. Light's article first appeared in *Technology and Culture*, the journal of the Society for the History of Technology, in 1999.

The archives of the Society of Women Engineers house the diary that is the basis of Lauren Kata's profile of Emma Barth written for *SWE Magazine* in 2002 and included here as Chapter 12. With degrees in German and education, Barth was unable to find a teaching job during the depression, and so turned to engineering. Starting out as a draftsman during World War II, Barth went to school at night for seven years to earn her engineering degree while working full time. The excerpts from her diary illustrate the day-to-day challenges of combining work and school, and her later activities to organize women engineers and encourage young women to pursue engineering careers.

Chapter 13 profiles Katharine Stinson, the first woman to earn an engineering degree from North Carolina State University (NC State), who was encouraged to pursue an engineering career when she met her hero, Amelia Earhart. Stinson earned her degree in 1941 after first being turned away when she tried to enroll at NC State, and spent her career at the Civil Aeronautics Administration (now the Federal Aviation Administration) working on aircraft safety. One of her first assignments was to convert light airplanes into gliders for pilot training during World War II. A licensed pilot herself, Stinson was also the third president of the Society of Women Engineers. Betsy Flagler profiled Stinson for the *NC State Magazine* in 1998, just a few years before her death.

While engineering professional societies were formed in the late nineteenth and early twentieth centuries, women were not accepted as members for many years, and it was not until the late twentieth and early twenty-first centuries that the first women were elected presidents of those organizations. The American Society of Civil Engineers, founded in 1852 and the first engineering society in the United States, elected Patricia Galloway, its first female president, in 2003. The nation's largest engineering professional society, the Institute of Electrical and Electronic Engineers USA, had its first female president, LeEarl Bryant, in 2001. In 2002 the American Society of Mechanical Engineers had its second female president in Susan Skemp, and Diane Dorland was the first female president of the American Institute of Chemical Engineers in the same year. Theresa Helmlinger became the first female president of the National Society of Professional Engineers in 2003. Passing over the many women engineers who blazed trails in the second half of the twentieth century and saving their stories for another day, I close this volume with an article I wrote for *SWE Magazine* in 2003 profiling these five women who served as the first or second female president of five of the largest U.S. engineering societies in the early years of the twenty-first century.

The chapters collected here illustrate the trials and triumphs of both notable and "average" women engineers during a time when any woman in an engineering job was considered unusual, and efforts by authors over the past 80 years to share and celebrate the ordinary and extraordinary lives of women in engineering.

4

1952   *Petticoats and Slide Rules*

Margaret Ingels, M.E.

THIS TALK WAS PRESENTED BY MISS INGELS SEPTEMBER 4, 1952, before the Western Society of Engineers. Miss Ingels is the first woman to graduate from the University of Kentucky's College of Engineering. *[See Chapter 10 for an early profile of Margaret Ingels. Ed.]*

In selecting the title for this talk, "Petticoats and Slide Rules," I planned to pay tribute to all women engineers who predate the "blue jeans" era. I thought at the time that the petticoat age of women engineers probably began during the days of cotton garments adorned with Hamburg ruffles. To my surprise, research proved that women entered the engineering profession during the multi-petticoat era; but we do not go back to hoop skirts and bustles!

The research to learn about early women engineers was an intriguing project, and seemed unending. The stack of information which came in from the secretaries of national societies, from the Women's Bureau, U.S. Department of Labor, from engineering colleges, and from men engineers grew higher and higher with each incoming mail. Soon it became evident that the complete story of the early women engineers was much too long to tell in the time allotted to me. Therefore, instead of recognizing all of them, my remarks will be limited to those who may be called "trail

Ingels, Margaret. (1952). "Petticoats and slide rules." *Midwest Engineer*, August, 2–4, 10–12, 15–16. Reprinted with permission of the Western Society of Engineers.

blazers" or "pioneers". And, for want of adequate data from foreign countries, this discussion will be limited to American women. I also fear I have inadequate data on American women, so I hasten to make an apology. I ask forgiveness of those women pioneers in the engineering profession who are unknowingly omitted because their names and works were not discovered during the study. I also ask forgiveness of those women who are recognized in this paper because they have advanced the frontiers for women engineers and thus earned the title of "pioneer" but who, by dynamic spirit, are much too young for me to pin such a label on them.

The data assembled in the historical study of early engineers in petticoats warranted chronological presentation. However, the procedure of dating women posed a problem. (It never occurred to me to use the year of birth, for those women still active in engineering, although many such dates are available from various editions of *Who's Who.*) After some consideration, the date selected was the time a woman entered the profession as a graduate with an engineering degree, the year she started work as an engineer, or the year she qualified as a member of an engineering society— the earliest of the three.

With the "ground rules" explained, and again with apologies for omissions, the procession of pioneer women engineers starts. They pass our reviewing stand.

1886: Edith Julia Griswold leads the procession of engineers in petticoats. To her goes the title of "Trail Blazer." She, born in 1863 in Connecticut, studied civil and mechanical engineering at the age of twenty-one, and for two years took a course in electrical engineering at New York City Normal College. She also studied law at New York University. In 1886 Miss Griswold opened an office on Lower Broadway in New York City as a draftsman, and specialized in patent office drawings. *Who's Who in Engineering* (1925) generously recognized her as an engineer and as a patent law expert. *[Ingels apparently did not discover Elizabeth Bragg, who graduated from the University of California, Berkeley, in 1876 with a civil engineering degree, in her research for this speech. Ed.]*

1893: The year 1893 brings Bertha Lamme before our reviewing stand. In that year she was graduated from Ohio State University and received the degree of M.E. in E.E. She accepted employment with Westinghouse Electric Company, which today lists a dozen women engaged in important engineering work and more than one hundred others in related positions. From 1893 to 1905, Miss Lamme worked on mathematics of machine design. She then married Russell Feicht, a fellow engineer in the Company, and retired from professional activity. Mrs. Bertha Lamme Feicht earned her position in the parade of women engineers as an early, if not the first, woman to receive a degree in engineering. *[Lamme is profiled in more detail in Chapter 8. Ed.]*

1894: Mrs. Lena Allen Stoiber enters the parade of petticoat engineers in 1894. In that year she qualified for associate membership in the American Society of Mining and Metallurgical Engineers as the owner of the Silverton Mine in Silverton, Nevada. An endorser of her application for membership in the Society recently recalled his acquaintance with her. He began his letter with, "Lena Allen Stoiber was a character—most intelligent—a masterful woman." He further adds that she asked him to manage her mine, but he declined as he felt that it would be "most unpleasant to be bossed by her." The endorser wrote that Mrs. Stoiber sold her mine, became socially prominent, with a fine house and ample wealth. In 1922 Mrs. Stoiber married again, and hyphenated her name to become Mrs. Lena Allen Stoiber-Ellis. She remained a member of the American Society of Mining and Metallurgical Engineers until her death in 1935.

1895: Next Marion Sara Parker passes before us. She was graduated from the University of Michigan in 1895 with a degree of B.S. (C.E.), having completed the four-year engineering course in three years. First she was employed in the Chicago office of Purdy & Henderson, structural engineers. When the firm opened an office in New York City, Miss Parker joined the staff there. The chief engineer in the New York City office recently recalled her excellent work. He wrote, "The complete calculations, foundation and framing plans for a large office building on Lower Broadway were put under her sole charge. When they were filed in the Building Department by the architect, to his complete and agreeable surprise, Miss Parker's plans came through with only three minor objections." He also wrote, "Miss Parker held her own with engineers and draftsmen in the office: her work was eminently satisfactory—neat, quick, accurate." About 1905 Miss Parker resigned, returned to her native state, and married, becoming Mrs. Marion Parker Madgwick.

1900: After Mrs. Madgwick walks Marshall Keiser. Her training and professional life earn her a unique position even among early women engineers. She was born in Alexandria, Kentucky in 1874, entered the Agriculture and Mining College of her native state university. Like another famous pioneer of Kentucky, Daniel Boone (who was a civil engineer), Miss Keiser blazed a trail. She continued her technical training at Ohio Mechanics Institute, and the Institute of Technology in Munich, Germany. After completing her formal education, she taught chemistry for five years. In 1900 she accepted a position as chemist of West Java Sugar Experimental Station on the Island of Java. In 1903 she became Mrs. Leland Wallace Holt, and retired from professional life until after her husband's death in 1908. She then became president and manager of the Holt Land & Cattle Company of Colorado and New Mexico, and president of the New Mexico Iron & Coal Mining Company of California. In 1912 she resigned from her executive positions and, from then on until her death in January, 1952, she engaged in farming, mining, and land development. She wrote many engineering articles which were published in chemical and mining journals. *Who's Who in America* records that Mrs. Marshall Keiser Holt adopted and educated six orphans—this in addition to her many activities in the engineering profession.

1903: Minette Ethelma Frankenberger joins the procession of women engineers in 1903 with a degree of B.S. (C.E.) from the University of Colorado. She worked as a draftsman from 1906 to 1909. Little is known of her professional career. However, we honor her as an early graduate in civil engineering.

1904: In 1904, Florence Hite joins the women engineers for a short period. After receiving the degree of civil engineering in architecture from Ohio State University, she married a mining engineer and did not continue in her professional career.

1905: In 1905, Nora Stanton Blatch, the first woman to be graduated from Cornell University with an engineering degree, enters the parade of pioneers—she with a degree in civil engineering. She is still marching forward with youthful vigor. She recently wrote that she had to put up with a "lot of razzing" when on campus, but many of the friendships made at the time have continued through the years.

When big company representatives arrived at Cornell in the spring of 1905 to select graduates, they did not discriminate against Miss Blatch—they could not as her grades were in the top five of her class, and had earned her membership in Sigma Xi. American Bridge Company employed her. She made good—became a "squad boss" after three weeks of employment.

Miss Blatch later became assistant chief engineer and chief draftsman, with thirty men working under her, at the Radley Steel Co. Because the wages of an engineer did not satisfy her, she established her own contracting business.

Today, Mrs. Morgan Barney, formerly Miss Nora Stanton Blatch, continues her contracting work in Greenwich, Connecticut, where she specializes in large residences—costing up to $135,000 each. *[Blatch was discussed in more detail in the previous chapter. Ed.]*

1914: There is a long gap in the procession of pioneer women engineers who entered the profession between 1905 and 1914, but there is no void. During this period Miss Kate Gleason made history for women engineers. In the school year of 1884–85, and again in 1888, she studied as a special student in Mechanical Arts at Cornell University. She was the first of the "Sibley Sues", a term applied to women enrolled in the Sibley College of Engineering of that great educational institution. She continued her studies in engineering with on-the-job training under her father at the Gleason Gear Works in Rochester. By 1914 her design of a worm and gear had brought her such distinction that the American Society of Mechanical Engineers accepted her as a member—the first woman to be so honored. She represented the Society several times at world conferences.

Miss Gleason made history in another profession not generally opened to women in her day. She was the first woman bank president in the country. She designed and manufactured fireproof houses with unskilled labor that won her additional recognition to that of being the first woman mechanical engineer. When asked in 1930, three years before her death, to what she attributed her success, she said, "A bold front, a willingness to risk more than the crowd, determination, some common sense, and plenty of hard work."

If entrance in the procession of pioneers were based on the date a woman decided to be an engineer, Kate Gleason would lead all others with her entrance in Cornell in 1884. However, by the "ground rules" she joins the parade of petticoat engineers in 1914 as a member of the American Society of Mechanical Engineers. In spite of the date, I think of her as a real "trail blazer", the leading spirit for all women engineers whose formula for success should be adopted by all in the engineering profession—both men and women. *[See Chapter 7 for more on Gleason. Ed.]*

1915: In August, 1914, Germany invaded Belgium. World War I started. Then, as in World War II, there was a shortage of men engineers. Women who aspired to enter the profession then, as do the daughters and granddaughters of their contemporaries today, found new doors opened to them. Alice Goff in her *Women Can Be Engineers* aptly describes the situation. She wrote, "By an irony of fate, war, always bitterly denounced by women, has advanced them in the engineering profession." *[Excerpts of Goff's book are presented in Chapter 10. Ed.]*

Among the first women to benefit from a shortage of engineers due to World War I were Alice C. Goff (she knew whereof she wrote) and Hazel Irene Quick. Both were graduated from the University of Michigan in 1915 with bachelor's degrees in civil engineering.

Alice C. Goff, after graduation, did appraisal work for the Dean of the College of Engineering for six months. She then joined Truscon Steel Company in Youngstown, Ohio, and has remained with the company ever since. For many years she was squad leader of a group of men. Today she designs and estimates reinforced concrete for large buildings in the United States and South America—one of her largest was a $15,000,000 bomber plant in Texas during World War II. She holds a professional engineer's license in Ohio.

Hazel Quick, after graduation in civil engineering in 1915, spent eighteen months in appraisal work and surveying. In 1917 she joined the Michigan Bell Telephone Company, where she became Assistant Plant Extension Engineer. Her work, until her retirement in 1950, concerned special services and introduction of newly developed telephone equipment. She is a charter member of the Detroit Chapter of the Michigan Society of Professional Engineers and is serving a second term (which expires March 1, 1955) as a mayor-appointed Commissioner on the City Plan Commission of Detroit. At her summer home, Miss Quick uses a tractor for gardening and operates her own electric power equipment for carpentry work.

Lydia G. Weld accompanies Miss Goff and Miss Quick, the 1915 entries. After attending Bryn Mawr, Miss Weld studied naval architecture at the Massachusetts Institute of Technology. She was graduated in 1903, and accepted a position with the Newport News Shipbuilding and Dry Dock Company in Newport News, Virginia. There she had charge of making finished plans for government ships. In 1915 she became an associate member of the American Society of Mechanical Engineers; in 1935 a full member. She retired in 1917 because of ill health, and moved to California. There she today runs a 400-acre alfalfa, hog and pear ranch. She maintains her interest in engineering and continues her membership in the American Society of Mechanical Engineers.

1916: In 1916, I join the procession of engineers in petticoats, with a degree in mechanical engineering from the University of Kentucky. *[The author is profiled in Chapter 10. Ed.]*

1917: Dorothy Tilden Hanchett, University of Michigan, B.S. (C.E.), 1917 next comes into view. She worked in the office of the City Engineer, Flint, Michigan for three years, then entered the teaching profession—still later she married.

Next, M. Elsa Gardner passes before us. In 1917 she became gauge examiner of the British Ministry of Munitions of War in the U.S.A. She had, by that time, earned a B.A. degree from St. Lawrence University (1916), and had begun study in machine design at Pratt Institute. She continued her formal training in engineering while employed as inspector of airplanes, airplane engines, and precision gauges used in aircraft. Her schooling included classes in engineering at New York University and the Massachusetts Institute of Technology. Elsa Gardner has written many articles on aerodynamics and has earned a high place in both active service in machine design and technical writing. She continues to bring credit to herself and to women engineers with her work in the Materials Division, United States Army Air Corps at Wright Field, Dayton, Ohio.

1918: In 1918, Dorothy Hall, University of Michigan, B.S. in Chemical Engineering, a Ph.D. in 1920, comes into view. For a short time she served as research chemist, then as Chief Chemist at the General Electric Company Laboratory in Schenectady. She then was married to Mr. Gerald R. Brophy and retired from the engineering profession.

Helen Innes comes into view in 1918, as in that year she proved her high status as a heating engineer when the American Society of Heating and Ventilating Engineers accepted her as an associate member—a full member in 1923. She combined study at the New York School of Heating and Ventilating and Pratt Institute with on-the-job training, and earned recognition with her wide and successful experience in the heating field. In the mid-twenties she became Mrs. J. A. Donnelly, retired from engineering to live at Largent Springs Manor, Largent, West Virginia until her death in 1935.

Marching with Helen Innes Donnelly is Doska Monical, a geologist from the University of California—1918. After graduation, Miss Monical accepted a position with Shell Company of California. Her academic standing and professional work won her membership in Phi Beta Kappa and Sigma Xi, and an associate membership in the American Institute of Mining and Metallurgical Engineers.

1919: Frances Martin Bayard, a "Sibley Sue" of Cornell University, Class of 1919, next joins the early women engineers. A severe attack of influenza in November of her junior year prevented her from completing her course, but it did not weaken her determination to be an engineer. She applied her two years' training, first as a draftsman with Sun Ship Building Company, then as a design engineer in her father's firm, M. L. Bayard & Company, Inc., of Philadelphia. In 1921 she married Dr. Harold A. Kazmann. Today she is a director in the company her father founded, which manufactures deck machinery for battleships, aircraft carriers, and destroyers. Recently Mrs. Frances Bayard Kazmann wrote "Mine may not be a success story (from an engineer's standpoint), but it has never been dull. Being a doctor's wife is a full-time job which included organizing and planning offices as practice expanded, and assisting in my family's business to help bring it back after the depression years to win Navy 'E's' during the War."

The year the next woman enters the procession of women engineers is difficult to determine. However, there is no doubt that it was early in the century. She, Edith Clarke, graduated from Vassar College in 1908, where she majored in mathematics and won a Phi Beta Kappa Key. She taught for two years, and then decided upon an engineering career. In the fall of 1911 she entered the University of Wisconsin as a sophomore in civil engineering. However, at the end of the year she accepted a position with the American Telephone & Telegraph Company and remained in the company until 1918 as a computer. During this period, she took night courses at Columbia University, and one in radio at Hunter College. In the fall of 1918 she entered Massachusetts Institute of Technology and, in 1919, received the degree of Master of Science in Electrical Engineering—the first woman to receive an electrical engineering degree from that institution. Again she taught—this time physics at Istanbul Women's College. In 1922 she joined General Electric Company in Schenectady. During her many years there, she accomplished much, wrote many articles on engineering subjects. So great have been her achievements that Tau Beta Pi Association awarded her the Woman's Badge—the only one given a woman for accomplishments in the field. Today Professor Edith Clarke teaches electrical engineering at the University of Texas, and continues to bring credit to women in the engineering profession. *[See Chapter 9 for a more extensive profile of Clarke. Ed.]*

The year 1919 brings Mrs. Arthur T. Edgecomb into the parade of pioneers—Mrs. Edgecomb, the former Hilda Counts, received her A.B. degree from the University of Colorado, taught high school mathematics and physics for two years; returned to the University, studied electrical engineering, and was awarded a B.S. in E.E. in 1919—the first electrical engineering degree granted a woman by the University of Colorado.

Mrs. Edgecomb recently gave a sidelight on the attitude towards women on campus in her college days. One girl was not accepted as an engineering student because she smoked and swore in a shocking manner. A show of hands of non-smokers among women engineers may be of interest at this point! Times have changed!

In 1919 Miss Counts joined Westinghouse Electric Corporation as one of thirty-two engineers out of over three hundred applicants for a student training course. After two years with the Company, she returned to Colorado for an E.E. degree. However, instead of getting a higher degree, she became Mrs. Arthur T. Edgecomb and retired from her engineering career until after her husband's death just prior to World War II. After fourteen years' retirement, Mrs. Hilda Counts Edgecomb re-entered the engineering profession. Today she is an electrical engineer on the staff of the Rural Electrification Administration in Washington, D.C.

The year 1919 also brings into view Mrs. Olive E. Frank. With night study and on-the-job training, Mrs. Frank combined engineering and executive talents to design heating equipment and to organize her own company, which prospered throughout her lifetime. In 1919 she joined the American Society of Heating and Ventilating Engineers; in 1927, the American Society of Mechanical Engineers. At the time of her death in 1946, she was President and Treasurer of Frank Heaters Company, Inc., of Buffalo and New York City.

1920: The year 1920 brings eight women into view and each has earned the right to hold her head high.

One is Ethel H. Bailey. She began work in the engineering profession after attending the Michigan State Automobile School in Detroit in 1918, and taking a special course at George Washington University in 1920. In *Who's Who in Engineering* (1925), the record of Miss Bailey's activities, which started as an inspector of airplanes and airplane engines with Cadillac Automobile Company in Detroit, depicts the early history of aviation. She worked on the Liberty-12 engines, the "Shenandoah", Type 12 bombers, and the T-3 transport plane. She was the first woman admitted to full membership in the Society of Automotive Engineers—this in 1920, and the American Society of Steel Treating. She has written many technical articles. Still young to be called a "pioneer," Ethel Bailey carries on her active career in engineering today at Massachusetts Institute of Technology as an administrator in the Department of Biology. The Department has four electron microscopes, X-ray diffraction apparatus, and other types of equipment, as well as an expertly staffed machine shop to keep the equipment in condition for biological research.

Olive Dennis, another of the 1920 entries, first earned a bachelor's degree in science and mathematics at Goucher College, then a master's at Columbia University. In 1920 she was graduated from Cornell University with a degree in civil engineering. She then joined the Baltimore & Ohio Railroad Company in Baltimore, Maryland, first as a draftsman in the Bridge Engineering Department, and then as a Research Engineer. From 1920 until her retirement in 1951, Miss Dennis earned many honors. She was the first woman member of the American Railway Engineering Association, and for many years served on its committee on the Economics of Railway Location and Operation. During World War II she was engineering consultant of the Division of Railroad Transportation of the Office of Defense Transportation. She holds a patent for a window ventilator for railroad cars and, to prove her versatility, a patent on Baltimore & Ohio's famous "Blue China" dinnerware.

When you ride the B & O, observe the excellent passenger service and equipment, the efficiency of operation, the charm of the cars' interior and furnishings—all a credit to Olive W. Dennis, a woman "railroading" engineer. *[See also Chapter 10 for more on Dennis. Ed.]*

Elsie Eaves, who was graduated from the University of Colorado in 1920 with a degree of B.S. in C.E, rates a "first" as she walks in the procession of women engineers—the first woman elected to full membership of the American Society of Civil Engineers. After spending two years with the Colorado State Highway Department, she joined the staff of *Engineering News-Record* (of the McGraw-Hill Publishing Company) in New York. Her attainments proved the value of possessing a talent rare in engineers—the ability to write. She is the exception to the rule which a past Dean of Engineering at Kentucky University told his students, "an engineer has more to tell than anyone, and tells it the poorest". Miss Eaves is a notable exception to that rule. Her contributions to engineering literature include a chapter in "Pulsebeat of Industry" and many articles in technical magazines. Her "Wanted: Women Engineers" has, no doubt, led many women into the field of engineering. *[See also Chapter 10 for more on Eaves. Ed.]*

Also in the 1920 entrees of women pioneers is the late Marie E. Luhring. Her career resulted from the shortage of men engineers at the end of World War I. Graduated from Hunter College with an artist's degree, she drew animated cartoons for awhile; then, in 1918, with twenty-five other women, she joined International Motors Company to be trained as an engineer—and to remain with the Company until her death in 1939.

Miss Luhring's aptitude for mechanical engineering prompted her to obtain further technical training at Cooper Union, from which she was graduated in 1922 with high honors.

To Miss Luhring goes the distinction of being elected an associate member of the American Society of Automotive Engineers in 1920. To her also goes the credit, as a key worker on the project, for the development of a gas-electric locomotive in the late 1920's. Her associates recall her as "a marvelous person, a marvelous worker".

Lou Alta Melton, a classmate of Elsie Eaves, walks in the 1920 group of early women engineers with a B.S. in C.E. degree from the University of Colorado. After employment as a draftsman and promotion to junior bridge engineer with the U.S. Bureau of Public Works, she married A. S. Merrill. Today Mrs. Lou Alta Melton Merrill lives in Missoula, Montana, and often uses her technical training to help out Montana State University by teaching in the Mathematics Department.

Dr. Mary Engle Pennington earns her place with the 1920 women engineers. In that year, she was elected a member of the American Society of Refrigerating Engineers to become one of its active and most outstanding members. Her entrance in engineering came after a long and successful career in physical chemistry and food research. Dr. Pennington earned her doctor's degree from the University of Pennsylvania in 1895, when she was twenty-three years old. Her work in foods led her into the mechanical refrigeration field. Here she contributed materially to improvements in the design of cold storage plants, precoolers, and commercial and household refrigerators. During the second World War, Dr. Pennington served the government with her wide experience on food preservation. She is a member of several technical societies—has received many awards. She represents the refrigerating field as she walks in the parade of petticoat engineers.

Nellie Scott Rogers enters the procession of women engineers in 1920 as an early member of the Society of Automotive Engineers. The late Miss Rogers earned her position with self-training, and deserves the recognition we give her. From 1922 to 1928 she served as President of the Bantam Ball Bearing Company, Bantam, Connecticut. She later served as active Vice-President and Treasurer until her retirement in 1929. She died in April, 1942.

Helen Smith walks with the other 1920 engineers with a degree of B.S. in E.E. from the University of Michigan. She combined the art of engineering with the profession usually associated with petticoats—home economics. She first worked for the Ediphone Company, then for many years for the Rochester Gas Company, then in the Home Economics College of Syracuse University. In each position she specialized in electrical appliances for the home. Helen Smith is now retired, and lives in Key West, Florida.

1921: Alice Gertrude Bryant, M.D., is one of several women who join the parade in 1921. She comes as a member of the American Society of Heating and Ventilating Engineers. With several other doctors at that time, Dr. Bryant was attracted to the profession by studies on air, conducted at the research laboratories of the society. Doctors joined with air conditioning engineers to determine the combination of temperature, humidity, and air motion which most people liked best. Dr. Bryant brought to the research, experience from a long and successful career in medicine. She studied at Massachusetts Institute of Technology and Women's Medical College, earning her M.D. degree in 1890. A pioneer in two professions which numbered few women at the turn of the century—medicine and engineering—she served both faithfully until her death in 1942, at the age of eighty.

Mildred Pfister, another 1921 entry, carries the title of "Power Plant Equipment Engineer" for a Cincinnati company. She qualified for the unique position with a bachelor of science degree—chemistry major—from the University of Cincinnati, from which she was graduated in 1919. Her work from then on proved she possessed unusual talents. From water conditioning and corrosion control, she moved to machine tool inspection—first for a large company and then for the Ordnance Department. She later joined the Cincinnati office of United States Engineers. And, all through that while, she did consulting work on the side. She holds a professional engineer's license in Ohio. Today she is spending full time as a consulting engineer in her home city, Cincinnati.

In 1921, two "Sibley Sues" from Cornell University, with degrees in mechanical engineering, join the parade of petticoat engineers. They received the first M.E. degrees issued to women by that institution. They are Marie Reith and Herma Marie Trostler.

Marie Reith joined the staff of the Consolidated Edison Company in New York City—is there today. She directs market analyses, investigates the substitution of alternating current for direct current, studies new uses for electrical equipment in the home, in business, in industry, and analyzes work procedures.

Unfortunately, no information could be obtained on Herma Marie Trostler.

Margaret Arronet, another "Sibley Sue", with a C.E. degree from Cornell University, walks with her 1921 M.E. classmates. She worked as a draftsman with the American Bridge Company, then as an assistant in the research laboratory of the Portland Cement Association. In 1922–23 she served on the Hoover Mission in Russia. In 1926 she married Franklin Nichols Corbin, Jr., and retired from professional life.

We have watched the parade of pioneer women engineers from 1886 through 1921. And still they come!

But this paper has grown too long. The review must end. Like observers of other types of processions, we grow tired of watching the marching of engineers in petticoats. However, like other observers of other parades, we linger on for some one of special interest to come into view. We wait to recognize Catherine Cleveland (Mrs. H. L.) Harelson, University of Kentucky, M.E., 1924, who, because she made the highest grades ever earned in the engineering college, was awarded the Tau Beta Pi Badge No. 1 for women engineers—this in 1934.

And, we continue to wait in our reviewing stand for another special person to appear—Dr. Lillian Gilbreth. She may well have joined in the procession of pioneer women engineers when she assisted her husband, the late Frank B. Gilbreth, in his standardization of construction practices which he started in 1904. She may, as a member of the staff of Gilbreth, Inc., have joined the procession in 1914, as at that time, she took full charge of the work for which she is so well known—time study, fatigue study, and skill study, that have been a boon to production. Management has long appreciated Dr. Gilbreth for her unique and valuable service to industry. Engineers recognized her contribution in their field in 1926, when the American Society of Mechanical Engineers elected her to full membership. So Dr. Lillian Gilbreth, with a bachelor's and master's degree in literature from the University of California, a doctor of philosophy degree from Brown University, enters the parade of pioneer women engineers as a member of A.S.M.E. in 1926. Engineers took further note of her work—the University of Michigan conferred on her an M.E. degree in 1928, and Rutgers College, a doctor of engineering degree in 1929. I could continue to tell you of the many works and the many honors that mark Dr. Gilbreth's long professional career—but you know of her achievements and of her high position among engineers. *[For more on Gilbreth, see Chapters 3 and 5. Ed.]*

When we think of the many women who entered the engineering profession in the distant and not so distant past, we are tempted to wait for others to come into view. However, time grows short; we leave the reviewing stand. But, the procession of women engineers continues on through the "roaring '20's," the depression years of the '30's, the war years of the '40's, and the procession will continue through the years which lie ahead.

The woman who joins the procession of engineers today, tomorrow, and tomorrow's tomorrow benefits by a rich heritage bequeathed to her by Edith Julia Griswold, Bertha Lamme Feicht, Marion Parker Madgwick, and Kate Gleason. She assumes automatically the responsibility to further prove that petticoats and slide rules are compatible, and she must not carry the responsibility lightly. Her task is to widen the trails blazed for her—and more. She must build them into great highways for women engineers of the future to travel, free of prejudices and discrimination. This she can do by following Kate Gleason's formula—apply hard work, courage, and plenty of common sense to her engineering job. At the same time, it would be well for her to expect no favors because she is a woman. The records prove she can succeed in spite of being a petticoat engineer.

5

Women in the Intellectual Development of Engineering

A Study in Persistence and Systems Thought

Martha Moore Trescott

TECHNOLOGICAL CHANGE AND ENGINEERING OF VARIOUS kinds have always been a part of human lives. The development of the modern engineering disciplines, linked as they are to the sciences and math, began to accelerate after the Renaissance. The construction of the Italian canals and hydraulic works in the fourteenth and fifteenth centuries, and the building of important bridges, roadbeds, and other civil engineering works in France within the next two centuries, gave much impetus to the rise of civil engineering as a scientific discipline. This branch of engineering was carried forward especially by French universities and by West Point in the United States. The coming of Newtonian

Trescott, Martha Moore. (1990). "Women in the intellectual development of engineering: A study in persistence and systems thought." *Women of science: Righting the record*, G. Kass-Simon and Patricia Farnes, eds., Indiana University Press, Bloomington, 147–165, 182–185. Reprinted with permission of Indiana University Press.

mechanics and the development of steam engines, particularly in England, laid the foundation for much modern mechanical engineering. Upon it professional groups such as the American Society of Mechanical Engineers (1880) were founded. Developments in the professionalization of civil and mechanical engineering led to similar developments in the field of electricity and chemical technologies in the nineteenth and early twentieth centuries. Professional groups of engineers in a variety of specialties, such as refrigeration, heating and ventilation, computer applications, and bioengineering, have subsequently continued to proliferate.

Throughout human history, women have contributed to technological development in major ways. Indeed, before the modern era, women and technology were not seen to be antithetical as they are today. As the various scientific fields became professionalized, however, women technologists, theorists, and inventors were not generally welcomed or included in the sciences. Engineering was no exception in this trend. Yet, even in this atmosphere, some women did persist, and many have made significant contributions, to engineering and to the other sciences and technologies as well.

There is much more substance to the history of women in engineering than historians of women and of science and technology have generally believed. It is true that in terms of numbers which form a "critical mass" women engineers have been an insignificant factor until the 1970s, having constituted one percent or less of the total engineering population of the United States. However, in terms of women's impact upon the theory and practice of their respective engineering fields historically, they have often had an inordinately great effect. Many women have become successful engineers and have contributed to the formulation of entirely new systems of thought and design, often establishing new paradigms in their areas. As women in this "ultramasculine" field, engineers like Lillian Gilbreth and Edith Clarke have faced far greater hurdles than those which have had to be faced by either men or women who entered other professions. These women therefore had to be particularly persistent, determined, shrewd, and intelligent. The 1970s represent an unprecedented phenomenon in the history of engineering in terms of the number of women entering engineering fields; this was largely due to a combination of factors, including the influence of affirmative action, the women's movement, and high salaries.

Since engineering deals with systems, those who can think in the context of systems can make special contributions to the evolution of engineering thought. Early women engineers seem to have been particularly able systems thinkers. So people such as Lillian Gilbreth, Edith Clarke [Chapter 9], Kate Gleason [Chapter 7], and Emily Roebling [Chapter 6] probably had much more impact on their fields than their small numbers would imply. These early women engineers invariably report that they were assigned a variety of leftover, odd jobs (even a good bit of "dirty work") which no one else wanted. Since it was difficult for women to obtain engineering jobs, they frequently took what they could get. The variety of menial tasks which these women performed year after year must surely have been onerous. But such assignments also meant that these women had to master a wide variety of areas in their fields, thus becoming very well acquainted with an entire engineering system and work environment. This wide-ranging, basic knowledge then served to shape their overview of a system of work, thought, and design.[1]

Because of the great barriers they had to overcome, women engineers seem to have been, in general, more persistent than their male counterparts.[2] Despite this, and even though they were often highly visible during their careers due to their extreme underrepresentation, they have not been adequately recorded, remembered, and assessed. Even the contributions of the greatest intellects among them have been buried in various ways. And, although the record and significance of the work of women in other professions, and in science particularly, have also frequently been lost to history, ascertaining the history of women in engineering may be especially problematical.

For one thing, engineers and technologists have historically left fewer written documents than have other scientists, since engineers have been more oriented to invention and design than to writing. For another thing, while modern science, medicine, and other professions have typically been very male-dominated, engineering has been extremely so. Of all professions, only the ministry and engineering have been so underrepresented historically by women. Until the 1970s women represented far less than 1 percent of all practicing engineers; by the early eighties they constituted only 2.5 to 3.0 percent of the total engineering labor force. In 1960, 4.2 percent of all physicists, 8.6 percent of chemists, 26.4 percent of mathematicians, and 26.7 percent of all biological scientists were women.[3] Third, and related to the second point, women who have become engineers and (or) who have contributed to the evolution of engineering theory and practice may well have been educated or trained in other fields. Until fairly recently, many women who have excelled in engineering were not "really" engineers, having been educated primarily in physics, chemistry, mathematics, psychology, education, and other areas. To cover adequately the history of women engineers in general, one must include women in engineering who have earned engineering degrees and those who have not, those who are licensed engineers and those who are not, those working in engineering but with degrees in other areas, and even those who "dropped out" of engineering work and who are or have been ostensibly working in other areas but who still use their engineering training in these positions.[4]

The lack of role models, mentors, and counseling and other information about schooling and jobs has been severe, and women have often entered engineering after many years or after careers in other areas (such as secretarial work and teaching) and often very indirectly. Many of those who lacked fathers, brothers, and other male relatives who were engineers have said that they just "fell" into engineering by a somewhat haphazard process.

In treating specifically the history of the intellectual contributions of women to engineering, one encounters very great problems. Written documents, particularly primary sources, are difficult to locate, especially in sufficient wealth to permit detailed assessment. Engineers and technologists often were not prolific writers, and they may not have thought their letters and notes particularly worth saving. Since their thoughts often pertained to the more pragmatic areas of life and since engineers necessarily must consult with people and work in the "real" world—even academic engineers—their efforts may at times have been recorded as part of a team and, therefore, are difficult to assess individually. Furthermore, women engineers were conditioned, along with other women, to be modest and to allow their contributions to be subsumed under a man's—in some cases, a relative's—name. And, in general, credit has not been forthcoming for women engineers from male peers.

Finally, in the intellectual history of women in engineering, a great and disparate variety of fields must be surveyed. One must include not only the basic engineering fields, such as electrical, mechanical, chemical, and civil, but also industrial, metallurgical, aeronautical, ceramic, marine, biomedical, general, and many other engineering specialties as well. Because the task is so formidable, and this chapter covers such a variety of fields, only the case study approach is feasible. Therefore, only a few examples from different areas of engineering will be cited. In many cases, these women have helped to institute new paradigms in their fields.

There are many omissions here, not only because of lack of space but also because of the burial of the achievements of women engineers, even in more recent decades. With ever-greater numbers of degreed women engineers working in all areas, with more women engineering Ph.D.'s than ever before, women engineers still win very few awards and honors from professional engineering societies. For instance, in 1980 the National Society of Professional Engineers commented, "Although NSPE has some awards for which women engineers would be eligible, to date no woman has received any of our awards. I hope we change that soon, but—."[5] The following accounts show just how the work of women in engineering has often been obscured.

Ellen Swallow Richards and
The Ecology Movement

Ellen Swallow Richards (1842–1911), who was denied an earned doctoral degree in chemistry at the Massachusetts Institute of Technology in the 1870s, has been called "the woman who founded ecology."[6] If this is a somewhat grandiose claim, it nevertheless reveals that she, like various other women in engineering, was a systems thinker.

Ellen Richards, although not an engineer by training, contributed much to the establishment of forerunners of environmental and sanitary engineering. She was associated with the MIT Chemistry Department from the time she earned a bachelor's degree there in 1873. From 1873 to 1878, she taught there without title or salary. In 1878 she became an instructor in chemistry and mineralogy in the Woman's Laboratory, which she helped found. From

Ellen Swallow Richards.
Photo courtesy of the MIT Museum.

Ellen Swallow Richards and MIT Faculty.
Photo courtesy of the MIT Museum.

1884 to 1911 she was the Instructor of Sanitary Chemistry at MIT. She performed numerous analyses of water and gas for the State of Massachusetts
and others and also became a specialist in analysis of metals and minerals.[7]

Ellen Swallow Richards authored more than fifteen books and numerous articles and reports, including *Home Sanitation, Cost of Living, Air,
Water and Food, Sanitation in Daily Life, Industrial Water Analysis*, and
Conservation by Sanitation.[8] It may not be a gross overstatement to call
her the founder of ecology. Certainly, she was a pioneer in the science and
engineering of the environment.

It is difficult to do justice to all the contributions Ellen Swallow Richards made to the understanding of the total human environment. In promoting concepts of environmental systems, she exhibited holistic thinking par
excellence. With a thorough grounding in chemistry, she began her study of
the environment with analysis of one of its most crucial aspects—water. And,
as Robert Clarke has noted, "water analysis was a new branch of science"
in the 1870s.[9] Ellen Swallow Richards was first introduced to it through
one of her professors at MIT, William R. Nichols (who, ironically, had not
believed in admitting women to MIT). As his best student, she performed

extensive analyses on Massachusetts sewage and water supplies under the auspices of the Massachusetts Board of Health in 1872, and Clarke has stated that this study "made her a preeminent international water scientist even before her graduation."[10] From the use of chemistry for the study of water, she then began analyses of the earth's minerals under another MIT professor and mineralogist and her future husband, Robert H. Richards. She distinguished herself also in this area, especially in her detection of small amounts of van[a]dium in various ore samples, a rare and very difficult metal to detect at that time. Although still a student at MIT, she was gaining a worldwide reputation for her chemical genius in environmental analyses. Her husband has written:

> My wife's work under Professor Nichols for the Massachusetts State Board of Health was arduous and extensive. Hundreds of analyses of water were almost entirely done by her. Much of the good work that was done later by Dr. Drown, whereby the standards of purity for wells, by means of curves called isochlors, was established for the whole State of Massachusetts, had its foundation in her work at this period. . . .
>
> My friend, David Browne of Coppercliff, Ontario, was seeking information about his copper ore from the Coppercliff Mine. He sent samples to a number of assayers, and among others, to Mrs. Richards. All the others returned results in copper and, I dare say, they did not know that they were to look for anything else. She, on the other hand, gave him a percent of copper in the ore and also reported five percent of nickel. This, I believe, was the beginning of the great nickel industry of which the Coppercliff Mine was the center. David Browne always said that Mrs. Richards was the best analyst in the United States.[11]

Her husband went on to report that in 1879 she was the first woman to be elected to the American Institute of Mining and Metallurgical Engineers, an honor which pleased her very much. (Incidentally, he was, as Margaret W. Rossiter has noted, vice-president of AIMME at the time.)[12]

In a file of her letters at MIT is correspondence which shows that Ellen Swallow Richards worked with both the Mining and Metallurgical Laboratory there and the Chemical Laboratory in performing analyses, as well as with the state board. Also in these letters one can see her concern for sanitation and the environment. She wrote Dr. Noyes, MIT president, about the state of heating and ventilation in certain MIT buildings in 1907–1909, saying that "one of the most serious problems of civilization is clean water and clean air, not only for ourselves but for the world."[13] Dr. Noyes responded that "it would seem that an investigation of it [ventilation of

MIT buildings] in detail with the purpose of discovering the most serious evils and suggesting practicable remedies for them would form a suitable and valuable thesis for one or two students in sanitary engineering or in the heating and ventilating option of the mechanical engineering course."[14]

She became an authority in analysis of food and the human diet, another vital aspect of the environment, and also on the subject of safer and healthier buildings and their design. She became so committed to pure air and water that she almost completely redesigned and remodeled the house which she and Robert Richards occupied in the Boston area, placing particular emphasis on its heating and ventilation systems. Clarke describes what she did:

> She checked and adjusted the plumb and fit of pipes through the house, replaced most with modern-seal joints, put in traps and other precautions for waste water, discarded the old lead poisoning water lines. A hand pump in the kitchen pulled water up from the well into a storage tank on the second floor for bath and toilet. This was before the municipal mains were laid in her neighborhood. She redesigned an inefficient water heater in the basement, replacing its input pipe and burner so that water would heat faster with less fuel. She put a "water back" on the furnace, using the heat from it in the winter and the water heater itself in the summer. . . . These devices became industry standards, but neither Robert nor Ellen tried to patent them. . . . Working with new knowledge of air analysis gained at the Institute, the Richards designed and installed a mechanical system of ventilation and circulation, a radical innovation in homes of any structure of their day.[15]

This was environmental engineering to an impressive degree. One major significance of this undertaking, aside from the many engineering innovations involved, is that this house became a consumers' testing laboratory, and she housed and taught students and others there. Some of her most important instruction actually occurred in the "Center for Right Living," as she called it.

Ellen Swallow Richards viewed her extensive work in the organization and education of women as a part of her environmental work. She felt that women, being the center of family life and therefore perhaps the most critical part of the human environment, needed to be educated about diet, adulteration of foods, proper ventilation of homes, and other features of home life to promote the health and safety of all. This viewpoint may appear, to a later age, to be advocating that women stay "in their place." However, no other woman in American history labored so tirelessly and successfully on behalf of women in science. To understand the extent of Swallow Richards's contributions, we must understand the context in which she lived and worked. There were virtually no opportunities for women to receive scientific education. The conditions under which most housewives lived were so unhealthy and unsanitary that they had to struggle to keep themselves and their families alive and even minimally free of disease.[16]

In 1876, the Woman's Laboratory at MIT, which Ellen Swallow Richards founded, introduced the first course in biology at MIT. She was very instrumental in establishing a life sciences curriculum there. This soon materialized into a full biology department. At that time, biology was not an acceptable area of instruction at schools of applied science. In 1892, at a public lecture in Boston, Ellen Swallow Richards called for the "christening of a new science"—"oekology."[17] The two main branches of this new science were the consumer-nutrition and the environment-education movements. As the twentieth century approached, Swallow Richards's interdisciplinary science of oekology fell victim to increasing specialization in the sciences. The study of environment became the province of the male-dominated life sciences (many of which she had helped to found and whose guiding lights she had taught). In this process environment became identified with plants and animals, and Swallow Richards's focus on the human environment was lost.[18]

The consumer-nutrition movement was carried forward more by women and by those untrained in science. Never a separatist in women's issues, and always identified in her own mind with the educated scientific community, Ellen Swallow Richards nevertheless found herself at the helm of this branch of oekology. It was not long before Swallow Richards's oekology was labeled "domestic science" and then "home economics." She was a founder of the American Home Economics Association and also of the New England Kitchen, one of the most innovative experiments ever undertaken in the daily, scientific preparation of wholesome foods for public consumption.

In addition, Swallow Richards continued to teach various courses at MIT, including sanitary engineering, a course which she introduced and first taught. In these classes, she educated "the men who went on to design and operate the world's first modern municipal sanitation facilities. 'Missionaries for a better world!' she lectured them."[19]

Ellen Swallow Richards also helped found the Seaside Laboratory, which later became the Marine Biological Laboratory at Woods Hole, where the sciences of both oceanography and limnology were developed.

But although Ellen Swallow Richards played a major role in the rise of these sciences, she has not been credited, as Robert Clarke recounts in some detail. Instead, men associated with the Massachusetts Board of Health and with MIT, many of whom she had taught and others with whom she had worked for years, were hailed as the "Father of Modern Sanitation" and the "Father of Public Health." But as Clarke comments, "if these were the 'fathers' of their individual fields, she was the 'mother' of them all."[20]

As a part of her early work in the consumer-nutrition movement, Swallow Richards had also been indirectly responsible for the design and introduction of the Aladdin Oven, which was a forerunner of today's ovens in home and industry. It was invented by the president of the Manufacturers Mutual Insurance Company, industrialist Edward Atkinson, whom she had met in conjunction with her survey on fire prevention in factories. This survey had, in turn, stemmed from an analysis she had undertaken of lubricating oils for factory machines and from the results of this analysis, which pointed to the combustible nature of many of these oils. She had then begun to develop noncombustible oils for machines to reduce the frequency of industrial fires. This work led her and Atkinson to the design of fire-resistant factories, which were "copied throughout industry"; much of the systems engineering was reminiscent of that seen in the remodeling of her own home. She ultimately became an authority on industrial and urban fires and began to survey schools and other public buildings.[21]

Her work, then, also involved much safety engineering as well as sanitation engineering, and she set up standards in many of the areas in which she worked. The development of isochlors, or the Normal Chlorine Map, which serves as "an early warning system for inland water pollution," is one of her lasting original contributions, along with the "world's first Water Purity Tables and . . . the first state water quality standards in the United States."[22]

Ellen was fortunate to be married to a fellow MIT scientist who openly praised her work. Hers is a case where marriage did not seem to hamper her work but perhaps enhanced it. From her student days until after her death, her husband did not seem jealous of her and, in fact, seemed proud of her and desirous of promoting her work among those he knew and in his writings. (However, their marriage in 1875 may help to account for MIT's unwillingness to pay her in the period 1875–1878.) Robert H. Richards also seemed to recognize her talents as a systems thinker. Perhaps his influence is one reason she has been as well recorded and remembered as she has been. Nonetheless, the extent of her work in founding many of the ecological sciences and related fields in engineering, and in the consequent introduction of entirely new paradigms, has not generally been recognized.

Lillian Moller Gilbreth and the Rise of Modern Industrial Engineering

Dr. Lillian Moller Gilbreth (1878–1972) is probably the best-known woman engineer in history. Perhaps the illumination of the "burial" of the most impressive intellectual contributions of "America's first lady of engineering" will set the stage for later discussions of less visible, less popular, or less famous women engineers.[23] By far, more primary sources, more archival materials and more published works are available for her life and work than for any other woman engineer.[24]

Lillian M. Gilbreth has certainly been considered important in the history of engineering. Her contributions to the rise of modern industrial engineering have been acknowledged in various ways by engineers and by

Lillian Gilbreth.
SWE Archives, Walter P. Reuther Library, Wayne State University, used with permission.

Lillian and Frank Gilbreth with 11 of their 12 children circa 1920s.
Photo provided by Purdue University Libraries, Archives and Special Collections, used with permission.

people outside the field of engineering. She "pioneered in the field of time-and-motion studies, showed companies how to improve management techniques and how to increase industrial efficiency and production by budgeting time and energy as well as money."[25]

In 1904, Lillian Moller married Frank Bunker Gilbreth, another pioneer in scientific management, who was especially noted for his very real genius in motion study. The Gilbreths worked together in many areas: in scientific management in their consulting firm (which advised many companies and became one of the most important firms in the United States); in research and writing (together they authored hundreds of documents); in lectures at various companies, universities, professional societies, and elsewhere; in conducting the Gilbreth summer schools on management topics; and in raising their twelve children (*Cheaper by the Dozen* is their story, written by a daughter and son).

Yet Lillian has received far less credit than has Frank from engineers and historians alike. She has been considered an adjunct to Frank, even though he died an untimely death in 1924, while she headed Gilbreth, Inc.

for decades afterward. She has been considered primarily his assistant or his disciple, even though he never earned a college degree and she attained the Ph.D. Her own expertise lay in the realm of integrating psychology and considerations of mental processes with time-and-motion work, while it is recognized among their colleagues that "if Frank Gilbreth slighted any discipline in his consideration, it was psychology."[26] If she is cited by historians, discussions are often limited to her contributions to domestic engineering (e.g., design of kitchen and appliances).[27] Lillian outlived Frank by nearly fifty years and was vigorous and professionally active almost until the time of her death in 1972. During this time she not only headed Gilbreth, Inc. (in effect she had been at its helm during much of Frank's lifetime, too) but also became a full professor of management in the School of Mechanical Engineering at Purdue University in 1935. She had succeeded Frank as lecturer there in 1924. She became head of the Department of Personnel Relations at Newark School of Engineering in 1941, and visiting professor of management at the University of Wisconsin at Madison in 1955. She received many honorary degrees (both master's and Ph.D.) in engineering from the 1920s on.[28]

In short, she deserves to be recalled and viewed in her own right, and not merely listed together with Frank as "the engineer, inventor, psychologist, educator Gilbreth."[29] While the first two terms may well describe Frank, Lillian was not only an engineer but also the psychologist and main educator of the couple. Despite the recognition and publicity of various kinds given Lillian Gilbreth in the past, the depth and breadth of her contribution to the establishment of modern industrial engineering have not been well understood or widely discussed by historians or by engineers. Her major contributions lie in two directions: (1) the incorporation of psychological considerations, as conceived in broad terms (problem solving and the behavior of individuals, and related topics such as incentives, the nature of the work environment, monotony, the transference of skill among jobs and industries, and so on), into time-and-motion thought and study; and (2) the establishment of industrial engineering curricula in engineering schools in this country and around the world.

With graduate studies in psychology at Brown (she obtained her Ph.D. in 1915), Lillian Gilbreth was perhaps the best-trained psychologist at the time who was also interested and working in time-and-motion study. Certainly neither Frederick W. Taylor, who is associated with innovations in time study and most often cited as the founder of scientific management, nor Frank Gilbreth had had any special training in psychology to enable them to analyze in a systematic and professional way areas dealing with psychology, work, and management. Since workers the world over often resisted the introduction of time-and-motion concepts quite strongly, it is doubtful that the work of either Taylor or Frank Gilbreth would have been wisely, if at all, accepted without the shrewd application of psychology to time-and-motion work. Lillian's insights and those of her students and other workers (both male and female) for whom she was a guide and authority in this area helped reduce workers' resistance. Such applications of psychology, along with Frank's work in physiological areas, helped establish the study of human factors in engineering design.

Of her many articles, books, reports, and lectures, her early book *The Psychology of Management,* which stemmed from her Ph.D. research, is the most important in the history of engineering thought. This book was termed by George Iles (who subsumed her work under Frank's name) as a "golden gift to industrial philosophy."[30] When the Society of Industrial Engineers made Lillian an honorary member in 1921 (she was the second person to be so honored; Herbert Hoover was also made an honorary member of SIE), it was commented that:

> she was the first to recognize that management is a problem of psychology and her book, *The Psychology of Management,* was the first to show this fact to both the managers and the psychologists. This book had a very small sale for two years after its publication but the demand has continually increased, until today it is recognized as authoritative.[31]

In the literature of scientific management before World War I, there was little coverage of such topics as the psychology of work and management, and it was in this relative vacuum at the early date of 1914 that *The Psychology of Management* appeared. It is true that others such as Hugo Munsterberg had studied industrial psychology at about the same time that Lillian Gilbreth began her research in psychology and management. However, as Robert T. Livingston (professor of industrial engineering at Columbia) commented in 1960, "Munsterberg's writings went largely unrecognized" for a long while. Furthermore, no one before Lillian Gilbreth in *The Psychology of Management* had brought together the basic elements of management theory, which are: (1) knowledge of individual behavior, (2) the theory of groups, (3) the theory of communication, and (4) a rational basis of decision-making.[32] Although not always using this modern terminology, Dr. Gilbreth dealt with all of these areas, some in more depth than others.

To the modern observer, "psychology" denotes a fairly well-prescribed area, but it is misleading to apply modern usage of that term to discuss the entire content of Dr. Gilbreth's book. Her subtitle, *The Function of the Mind in Determining, Teaching and Installing Methods of Least Waste,* more nearly captures the scope. Hers certainly is not primarily a concern with the field of "industrial psychology," as both contemporary and modern psychologists and others might have understood that field. Indeed, she was writing in the context of scientific management, which had previously focused mostly on the physiological rather than the mental and emotional characteristics of workers and managers. Her book analyzes in detail the "function of the mind," or problem solving, decision making, planning, communicating, measuring, and evaluating in various work and managerial environments.

Throughout *The Psychology of Management*, she is able to transform time-and-motion study into the rudiments of modern managerial practices. In setting forth the value of her approach, she draws on the literature available, including her husband's work, but it is clear that her analysis represents a new point of departure in management. Her insights into the process of disciplining the work force, for example, and her analysis of the inherent conflicts of interest if disciplinarian and foreman are one and the same person show simultaneously the stark contrast not only between her ideas and the approach of "traditional," older ways of management but also between her approach and that of Taylor and his male disciples. She indicates that the foreman or disciplinarian must consider not only a specific act committed by a worker but also the history of this worker's behavior, his or her physical condition, the relative effectiveness of different kinds of and settings for discipline for a particular person, the sensitivity of the person being disciplined, identification of any ringleader and the necessity of disciplining that person too, and so forth. And she then says that the words "disciplinarian" and "punishment" as employed by scientific management "are most unfortunate."

> The "Disciplinarian" would be far better called the "peacemaker," and the "punishment" by some such word as the "adjustment." . . . The aim is, not to put the man down, but to keep him up to his standard, as will be shown later in a chapter on Incentives.[33]

Her analysis here also demonstrates her empathy with how workers *feel*. There was little in the literature of scientific management at that time about sensitivity to workers' feelings, perhaps, ironically, since labor often felt very strongly opposed to the new management. In that area, she was very far ahead of her time, since the rise of sensitivity training in management has really only blossomed since the 1960s.

From her chapter on individuality it is easy to see her pioneering efforts in management theory, especially in "knowledge of individual behavior."

And yet not even this most obvious contribution is noted in the American Society of Mechanical Engineers volume covering *Fifty Years Progress in Management*, 1910–1960. Munsterberg, Gillespie, Lecky, and other psychologists are cited but not Lillian Gilbreth (who, incidentally, was co-author of the introductory, overview essay).[34]

In 1911 Lillian Gilbreth introduced the first mention of the psychology of management at any management meeting, at the Dartmouth College Conference on Scientific Management.[35] In 1924 Lillian commented on a paper by H. S. Person, "Industrial Psychology," delivered at the Taylor Society. In referring to the Dartmouth conference, she said, "It is now almost thirteen years since the importance of [the relationship between psychology and management] was stressed before those interested in scientific management. . . . For seven years before that time, steady progress had been made in correlating psychology and management, but from that time on the correlation was placed upon a scientific basis."[36] Recall that Lillian had received her Ph.D. in psychology in 1915 and had married Frank in 1904. (It is interesting that Frank's many publications are typically post-1904.) Since this work formed the basis for her dissertation,[37] which she was working on before 1915, it is clear that she had been among those studying the interfaces between psychology and management—both in industry and in the home—during the "seven years" before 1911. It is not an overstatement to term her a foremost pioneer in this area in the earliest days of such work.

In the Gilbreth Collection at Purdue, it is interesting to note a brief career sketch of her work, differentiating her contributions from Frank's. (She wrote this as part of her application for membership in the American Society of Mechanical Engineers (ASME), which had not exactly sought her as a member, despite her renown.)[38] Between 1904 and 1914, when Frank B. Gilbreth, Inc. "operated as a construction company," she said, "I was chiefly employed in the systems work, standardizing practice. The results were published in *Field System, Concrete System*, and *Bricklaying System*"—all of which show only Frank's authorship. She continued:

> I was also engaged in the perfecting of the methods and devices for laying brick by the packet method, and in the design and construction of reinforced concrete work. This work had to do with the management as well as the operation end. . . .
>
> 1914–1924. In 1914 our company began to specialize in management work. I was placed in charge of the correlation of engineering and management psychology, and became an active member of the staff making visits to the plants systematized in order to lay out

the method of attack on the problems, being responsible for getting the necessary material for the installation into shape, working up the data as they accumulated, and drafting the interim and final reports. I was also in charge of research and teaching, and of working up such mechanisms, forms and methods as were needed for our type of installation of scientific management, motion study, fatigue study and skill study. These had to do not only with the handling of men, but with the simplification and standardization of the machinery and tools, for the use of both the normal and the handicapped. During Mr. Gilbreth's frequent and prolonged absences both in this country and abroad, I was in responsible charge of all branches of the work. This was also the case while he was in the service, and while he was recovering from his long illness incurred therein.

1924. Since Mr. Gilbreth's death, June 14, 1924, I have been the head of our organization, which consisted of Consulting Engineers and does work in management, and have had responsible charge of the research, installation and the teaching, in this country and abroad.[39]

As both Lillian and Frank conceived the rise of scientific management, teaching was integral for implementing and disseminating its practice. It was new and unfamiliar and was sometimes resisted: it had to be taught within firms, in the universities, and to others. Lillian Gilbreth was a brilliant teacher as well as a researcher. She had a grasp of theory and practice, and knowledge of the evolution of the field, and could teach the methods of scientific management to workers and managers. Many of her colleagues noted she had great tact and diplomacy, which served her well in integrating developments in industry and the universities.

Her upper-level undergraduate and graduate courses in management at Purdue in the 1930s and the 1940s were "open only to graduate students and to seniors of outstanding ability" and covered "investigation of specific management problems in the fields of organization, time and motion study, industrial accounting, factory layout, economic selection and equipment, and similar topics." By the 1940s she was listed with the faculty of industrial engineering at Purdue, whereas earlier in the Purdue catalogues her courses had been listed with general engineering or with industrial management.[40]

She was an invited speaker on numerous college campuses throughout her life, and addressed not only engineering students, but also at times women students specifically. She authored papers encouraging women to go into industrial engineering and management in this country and abroad, and at the Gilbreth summer schools, which she conducted, at least half the participants from various countries (mostly European) were women.[41]

Indeed, Lillian's work with and on behalf of women—from the handicapped homemaker to the female worker in the factory and office, to the professional in management and engineering, and to women consumers in general—has not begun to be illuminated. Her efforts on behalf of women in various roles and jobs and in different social and economic classes are a forgotten chapter in women's history.

Certainly, not only her work with and for women but also her various contributions to industrial engineering and its precursors, spanning nearly seven decades, have been only partly and vaguely acknowledged. She, in fact, helped formulate much of the theoretical underpinnings of the field, but she has been too narrowly labeled a "psychologist."[42] The precursors of the modern notion of "the work of a professional manager," in terms of "planning, organizing, integrating and measuring," as in Harold Smiddy's conception, can be seen in Lillian's published works, including *The Psychology of Management*, and in the records of studies she did for various firms, held in the Gilbreth Library. Yet when Smiddy and others today view the evolution of ideas about the function of the mind in management, Lillian Gilbreth's pioneering work is typically not mentioned.[43]

After Frank's death in 1924, she continued to be a prolific writer and to participate in meetings of professional groups such as the ASME, presiding over sessions such as one on the "management researcher" in 1933.[44] Even before her husband died, the Gilbreths together authored well over fifty papers on scientific management topics, not including those written by each of them as sole author nor including their books and consulting reports. And while Lillian's name alone appears on a few of these papers, it is not difficult to suppose that she was the main author on at least such works as "The Place of the Psychologist in Industry," "The Individual in Modern Management," "Psychiatry and Management," "The Relation of Posture to Fatigue of Women in Industry," and others.[45] She may well have been the principal investigator and author on articles credited to them both in the areas of fatigue, standardization, and transference of skill, but that is difficult to determine. The fact is that her own originality was buried, not only because she was married to a man in the same general field but also because of Frank's wish that both their names appear on all they wrote, even though this did not always happen.[46] Yet, because her own expertise and that of her husband were so clearly differentiated in many areas, and because she outlived him long enough to establish authority "in her own right," it is possible, at least in part, to resurrect her unique contributions.[47]

It is important to remember the context of Lillian's work, which emphasized the human element in scientific management. Being among the first to be so concerned with the human factor meant that one undertook to explore a frontier. Many subject areas were legitimate aspects of her research, since the human element is pervasive in all areas of work. As she and others who came after her worked, an increasing number of avenues for investigation were opened and a certain definition of the field evolved. *The Psychology of Management* was a wedge, opening whole new areas to scientific management, which have later evolved into mainstream topics in industrial engineering. *The Psychology of Management* was a departure from classical scientific management and formed a basis for much modern management theory.

Because of her exceptional longevity, her creativity and productivity in her consulting work, her research and publications, her lectures, courses, and workshops all over the world, and, therefore, her prestige and popularity, Lillian Gilbreth may have contributed more in the first four decades of this century than any other single person to defining industrial engineering and its major areas of investigation and analysis. Yet, even though she did remain very active for a long time and was a prolific writer, it is alarming that history has ignored many of the most significant contributions of Lillian Moller Gilbreth, "Member No. 1" of the Society of Women Engineers and perhaps the foremost woman engineer in history.[48]

[. . .]

It has been and remains difficult for women to become recognized as authorities in engineering. This will undoubtedly abate somewhat as the old myths and stereotypes about women and their supposedly inferior mental capacities, especially in technical fields, are being put solidly to rest. Such prejudice has been particularly unfortunate as applied not only to all women engineers but especially to those among them who have made substantive intellectual contributions to the theory and practice of engineering. Those women who combined the characteristics of pioneer, thinker, engineer, and woman in our society—many breaking new ground—may well have represented a highly select group of systems thinkers. They therefore may be reassessed as having contributed much more to technological change and the advancement of engineering knowledge than one might have supposed from their relatively small numbers.

Notes

1. Martha M. Trescott, *New Images, New Paths: A History of Women Engineers in the United States, 1850–1980. [Privately published, copyright 1996 by the author. Ed.]*

2. This is a truism in the engineering literature and is also overwhelmingly a response on the returned questionnaires (see note 1, above). Also see, e.g., Ellis Rubinstein, "Profiles in Persistence," *IEEE Spectrum*, November 1973, pp. 52–64.

3. Carolyn Cummings Perruci, *The Female Engineer and Scientist: Factors Associated with the Pursuit of a Professional Career* (West Lafayette, IN: Purdue University, 1968), pp. 1–2.

4. This is the scope of the author's current research project on the history of women engineers in the United States.

5. Letter, Jean Robertson, Director of Information Services, National Society of Professional Engineers, to Martha M. Trescott, July 9, 1980.

6. Robert Clarke, *Ellen Swallow, the Woman Who Founded Ecology* (Chicago: Follett, 1973). See also Robert H. Richards, *Robert Hallowell Richards, His Mark* (Boston: Little, Brown, 1936) p. 153.

7. Richards, *Robert Hallowell Richards, His Mark*, p. 170.

8. Lists of published writings of Mrs. Richards in MIT Archives.

9. Clarke, *Ellen Swallow*, p. 38.

10. Ibid., p. 39.

11. Richards, *Robert Hallowell Richards, His Mark*, pp. 159–160.

12. Ibid., p. 161. Also Margaret W. Rossiter, *Women Scientists in America, Struggles and Strategies to 1940* (Baltimore: The Johns Hopkins University Press, 1982), p. 91.

13. Letter, Ellen H. Swallow Richards to Dr. Noyes, President of MIT, n.d. (in December 1907), MIT Archives, file on Ellen Richards.

14. Letter, Dr. Noyes to Ellen Swallow Richards, January 22, 1909 (they had had correspondence on this question during 1907–1909).

15. Clarke, *Ellen Swallow*, pp. 66–67.

16. Ibid., see especially chapter 10.

17. Ibid., see especially chapter 12.

18. Ibid., see especially chapters 13–15.

19. Ibid., p. 141.

20. Ibid., p. 149

21. Ibid., pp. 122–125.

22. Ibid., p. 147.

23. Anon., "Lillian Moller Gilbreth: Remarkable First Lady of Engineering," *Society of Women Engineers Newsletter*, XXV (November/December 1978), 1.

24. Gilbreth Collection, Purdue University, West Lafayette, Indiana. See acknowledgment above.

25. "Lillian Moller Gilbreth," *SWE Newsletter*, p. 1.

26. Discussion by William G. Caples, American Society of Mechanical Engineers, *The Frank Gilbreth Centennial* (New York: ASME, 1969), p. 72.

27. For example, see Siegfried Giedion, *Mechanization Takes Command, A Contribution to Anonymous History* (New York: Norton, 1969), pp. 121, 525, 615–616.

28. "Lillian Moller Gilbreth," *SWE Newsletter*, pp. 1–2, and Dr. Gilbreth's resume, located in the biography file, Schlesinger Library.

29. See the outline on the History of Scientific Management in NHZ 0830-23, Gilbreth Collection, Purdue. It is evident that Lillian, as with Julia B. Hall (see Martha M. Trescott, "Julia B. Hall and Aluminum," *Dynamos and Virgins Revisited: Women and Technological Change in History*, ed. Trescott (Metuchen, NJ: Scarecrow Press, 1979), pp. 149–179), cooperated in allowing her work to be subsumed under Frank's name. This was probably due to her generally self-effacing nature and to her conditioning that men should get most of the credit while women should feel content and privileged to be considered their assistants. Also see, for example, the ASME Fifty Years Progress index, where all of the Gilbreths' work is listed together under "Gilbreth, Frank B., and Lillian M.," even though the text referred to may only mention one or the other (usually Frank).

30. George Iles, Introduction to Frank B. Gilbreth and L. M. Gilbreth, *Applied Motion Study, A Collection of Papers on the Efficient Method to Industrial Preparedness* (New York: Sturgis & Walton 1917), p. xi. Under Frank's name here are given his title, "Consulting Management Engineer," and a list of his professional society memberships, whereas with Lillian's initials only "Ph.D." is noted.

31. Anon., "Honorary Member No. 2," *Society of Industrial Engineers Bulletin*, III (May 1921), 2–3, Gilbreth Collection.

32. American Society of Mechanical Engineers, *Fifty Years Progress in Management, 1910–1960* (New York: ASME, 1960), p. 126.

33. Lillian M. Gilbreth, *The Psychology of Management: The Function of the Mind in Determining, Teaching and Installing Methods of Least Waste* (New York: Sturgis & Walton, 1914), p. 72.

34. ASME, *Gilbreth Centennial*, especially p. 126.

35. Item labeled "Dartmouth College Conference" (NHZ 0830-23, with typewritten note by LMG, 2113-41, saying (in reference to her remarks at the Dartmouth Conference in 1911), "This is the first mention of the Psychology of Management [sic] at any management meeting." Gilbreth Collection, Purdue University.

36. Typescript of Lillian M. Gilbreth, "Discussion of Dr. Person's paper—Industrial Psychology, Taylor Society, Boston, Mass., April 25, 1924," p. 1, found in the Gilbreth Collection (NHZ 0830-11).

37. See especially Nancy Z. Reynolds, "Dr. Lillian Moller Gilbreth, 1878–1972," *Industrial Engineering*, February 1972, p. 30.

38. See letters to and from Mrs. Gilbreth on her election to ASME membership during 1925–1926 in the file NHZ 0830-1, Gilbreth Collection. Had she been a man in the field with such outstanding qualifications and publications, she would have undoubtedly been sought for membership and honored, instead of having to go to such lengths. This is another fascinating story.

39. Lillian M. Gilbreth, biographical memo, 1926, in NHZ 0830-1 Gilbreth Collection.

40. From the *Bulletin of Purdue University*, 1934/35–1949/50, as found in a search of these university catalogues by Keith Dowden.

41. See letter, W. H. Faunce to Lillian Gilbreth, May 7, 1928, Gilbreth Collection, NHZ 0830-4, and letter, Richard L. Anthony to Lillian Gilbreth, February 26, 1941. Also see letters, Arnaud C. Marts to Lillian Gilbreth, December 12, 1940 and January 8, 1941, and letter, Dorothy Dyer to Lillian Gilbreth, February 13, 1941, Gilbreth Collection, NHZ 0830-5. Also, "Stevens Honors Five Engineers," *The Christian Science Monitor* (n.d., n.p.), clipping from Gilbreth biography file, Schlesinger Library. An inkling of the number of speeches and her extensive travel through the decades can be seen in the index to boxes of material on Lillian Gilbreth alone, Gilbreth Collection, Purdue. Materials on the Gilbreth summer schools, particularly the one held in Baveno, Italy, June 10–25, 1927, Gilbreth Collection NHZ 0830-31. Lillian M. Gilbreth, "Opportunities for Women in Industrial Engineering," Mimeo, October 20, 1924, pp. 2–3, Gilbreth Collection, NHZ 083-60. On women see also letter, Lillian Gilbreth to George C. Dent, September 19, 1924, and letter, George C. Dent to Lillian Gilbreth, October 1, 1924, containing a cover letter (sent with questionnaires) to all SIE women members, Gilbreth Collection, NHZ 0830-42. Letter, Dent to L. M. Gilbreth, October 1, 1924; also see in this same file the seven questionnaire responses (besides Mrs. Gilbreth's), and form letter to the "Women Members of the Society of Industrial Engineers" from George C. Dent, November 29, 1924, same file as noted above. Also see ASME, *Gilbreth Centennial*, p. 109, and ASME, *Fifty Years Progress*, pp. 136, 138–139. In addition, refer to Catherine Pilune, "The Industrial Engineer," mimeo, p. 1, Gilbreth Collection, NHZ 0830-112, and Gilbreth, "Opportunities for Women in Industrial Engineering," pp. 1–2. See also Lillian M. Gilbreth, "Industrial Engineering as a Career for Women," mimeo, n.d., Gilbreth Collection, NHZ 0830-67.

42. Dr. Gilbreth did treat such topics as psychologists typically do. See, for example, "Possible Psychopathic Types in Industry," a one-page typescript, n.d., NAPEGTG 0099, listing twelve types, attached to a one-page manuscript, in her handwriting, entitled "Psychiatry and Management." However, her main contributions should not be narrowly construed as those of a psychologist.

43. Harold F. Smiddy, "Management as a Profession," ASME, *Fifty Years Progress*, pp. 26–41.

44. Caples, in *The Frank Gilbreth Centennial*.

45. Ibid.

46. Letter, Frank B. Gilbreth to Irene M. Witte, as cited in the ASME, *Gilbreth Centennial*, p. 107.

47. Historians have, in fact, not covered either of the Gilbreths well. For some mention of them, see David F. Noble, *America by Design: Science, Technology, and the Rise of Corporate Capitalism* (New York: Knopf, 1977), especially pp. 274–275; Melvin Kranzberg and Carroll W. Pursell, Jr., ed., *Technology in Western*

Civilization (New York: Oxford University Press, 1967): Richard S. Kirby, Sidney Withington, Arthur B. Darling, and Frederic G. Kilgour, *Engineering in History* (New York: McGraw-Hill, 1956); David S. Landes, *The Unbound Prometheus*, (Cambridge: Cambridge University Press, 1969); Eugene S. Ferguson, *Bibliography of the History of Technology* (Cambridge, MA: MIT Press, 1968), pp. 301–303 and also 116–117; and Alfred D. Chandler, *The Visible Hand: The Managerial Revolution in American Business* (Cambridge, MA: Belknap Press, 1977), p. 466.

48. "Lillian Moller Gilbreth," *SWE Newsletter*, p. 1.

6

1984   *Emily W. Roebling*

One of the Builders
of the Bridge

Alva T. Matthews

"THE BROOKLYN BRIDGE WAS BUILT BY A WOMAN, THERE was a woman in charge." "A woman was the engineer of the bridge—the real mind behind the construction." Such rumors about the role of Emily Roebling abounded during the last six years of the building of the bridge. The fact that her husband, Washington Roebling, the Chief Engineer of the bridge, disappeared from public view with a nervous disorder that kept him from any public contact, even with the assistant engineers, led to this speculation. Before I discuss the role of Emily Roebling in the bridge's construction let me cite the text of a plaque that sits on one of the towers:

> The Builders of the Bridge: Dedicated to the memory of Emily Warren Roebling, with faith and courage she helped her stricken husband Colonel Washington Roebling to complete the construction of the bridge from the plans of his father, John A. Roebling, who gave his life to the bridge. Back of every great work we can find the self-sacrificing devotion of a woman.

Matthews, Alva T. (1984). "Emily Roebling, one of the builders of the bridge." *Bridge to the future: The centennial celebration of the Brooklyn Bridge*, Margaret Latimer, Brooke Hindle, and Melvin Krantzberg, eds., © New York Academy of Sciences, New York, N.Y., 63–70. Reprinted with permission of Blackwell Publishers Ltd.

Emily Roebling.
Special Collections and University Archives, Rutgers University Libraries, used with permission.

This is probably a very accurate statement about her role, and one that she herself would have approved of. But I cannot help wishing that the last statement had been a bit different, to give the idea that she was indeed one of the builders of the bridge, an integral part of the great work, more than a noble support.

Emily Roebling *[1843–1903. Ed.]* was born and grew up in the Hudson Valley, in Cold Spring, New York. Her family was not wealthy, but they were considered gentry. She was well-mannered, well-spoken, orderly, and well-educated for a young girl of her time. Significantly, she took all the mathematics available in the Cold Spring schools she attended. One of the strongest influences in her life was her brother, General G. K. Warren. A graduate of West Point, he was an assistant professor of mathematics there for a while, and then an engineer with a particular interest in bridges. He served as a general in the Union Army during the Civil War, and it was through him that Emily met Washington Roebling. In fact, her future husband and her brother possessed very similar traits. They both were free of pretense, had little fear of physical danger, and possessed great perseverance. The General fostered Emily's scientific bent by encouraging her in one of his hobbies—botany. Emily plainly adored him. But her closeness to her brother, when he was around, probably prepared her in some subtle way for the engineering mind, the engineering approach to life, with which she would be so intimately connected.

Washington Roebling was an aide to General Warren during the Civil War, and was introduced to Emily by the General's wife. They were engaged six weeks after the meeting and married the following year, in January 1865, when he left the army as a Lieutenant Colonel. Emily was twenty-one years old, and her husband was almost nine years her senior. Bridges were immediately at the center of their lives, for they spent the first two years of their marriage in Cincinnati with John Roebling, building the Ohio River Bridge. The Roeblings' first (and only) child was born in Germany in 1867 while they were touring Europe so that Washington could study the European advances in pneumatic caissons. This tour was at the request of John Roebling, who wished his son to make such a study in preparation for the design of the Brooklyn Bridge.

When John Roebling died in 1869, his young son, Washington, was already well versed in bridge building and extraordinarily knowledgeable in the design of pneumatic caissons. With some concern, the Board of Trustees of the bridge chose Washington to continue the work. Emily was then 25 and had a small son to care for.

In the spring of 1870 work began inside the Brooklyn caisson. These caissons, inverted boxes which would move down through the river bottom to solid footing, were potential torture chambers to those working inside. The internal air pressure increased with every inch of progress downward. The men doing the excavating experienced all manner of discomfort with the increasing pressure, but no one really knew why. At a depth of 45 feet, where the Brooklyn caisson hit rock, some of the men felt some of the symptoms— intense pains, violent cramps, nausea, dizziness. The caissons on the New York side, however, would have to go to 78½ feet before stopping. As that depth was approached, more and more men were affected, some so severely that death resulted. Washington Roebling collapsed with such symptoms in June of 1872.

For weeks Emily saw her husband near death, withstanding the pain only with the help of morphine. Through the summer and into the fall, the attacks recurred. The public was not told. Roebling went to work sporadically, and they took a brief trip, but by September he was staying home two and three days a week, and by December he could not leave their home at all. During the whole winter, in spite of acute suffering, Washington Roebling undertook to write in the most minute detail, the instructions and specifications for the work still to be done. The writing became agony, and talks with the assistant engineers became torture. In April of 1873 a formal leave of absence was requested and husband and wife went to Germany for a rest, quite possibly hoping for a miracle. They stayed six months, much longer than expected, and returned late in the year. Still not well, Washington was moved to Trenton, sixty miles from the bridge, where they set up their household and stayed for three years. "[T]he entire time the towers were being finished, the anchorages built, the cable-making machinery assembled and set in position, the Chief Engineer was nowhere near the bridge, and could see nothing of it." (McCullough 1972, p. 340) Under these circumstances, Emily, who had been his nurse, now gradually became his private secretary as well.

All the instructions to the bridge were sent by mail, including the complete specifications for the stonework on the New York anchorage and the wire rope for the footbridge. He would dictate a draft to Emily, correct it, and then the final copy was made. All of this was done in her longhand. It was a slow, tedious procedure. But at that pace, a scientific, alert mind such as Emily's, although unschooled in technical matters, must have absorbed an enormous amount. She probably first grasped the terminology, some of which was probably familiar to her already, and then, more slowly, the reasons why things were done and the scientific principles behind them. Washington Roebling in all his writing used language that was "patient, plain and to the point." (McCullough 1972, p. 340) There is no reason to believe that he failed to talk that way, too. In addition to handling his voluminous correspondence, Emily also read the incoming mail to him and all the newspapers she could. By filtering all this incoming and outgoing information through herself she gained a thorough overview of all the bridge's problems.

The competency of the assistant engineers during the absence of the Chief Engineer was extraordinary. While Washington was in Europe, C. C. Martin became the overall supervisor, and the work proceeded very well. But all the assistant engineers, with the exception of Farrington, were new to the building of suspension bridges. For all their competence, the work could not have continued without the detailed instructions of Roebling. That meant that the avenues of communication between Roebling and his assistants had to be open and clear, which is why Emily's role began to be so crucial. At first her secretarial work required efficiency and organization. But later, when Roebling could not handle the personal interviews with the assistants, her ability to communicate the information with enough assurance to create confidence in what she was saying was absolutely necessary to the success of this project with an absentee chief.

About this same time matters for her brother were worsening. He had long sought a court of inquiry to clear his name from the Five Forks incident, a Civil War battle during which he had been relieved of command. The whole affair had slowly eaten away at General Warren emotionally, while his physical health seemed to have been ruined by responsibility for construction of the Rock Island Bridge over the Mississippi. Emily found herself witnessing the gradual destruction of the two men who meant the most in her life. She had a young son to bring up, a husband near death many times, and a beloved brother being slowly destroyed. But she had the stamina of youth, and quite possibly all the work for the bridge, which required her full attention, helped her to withstand the pressures. It is possible that she deliberately threw herself into that work, as tedious as some of it must have been, to avoid some of the other realities she had to face.

There were political problems as well—unfounded charges of collusion with a supplier and a lawsuit for infringement of one aspect of the caisson design. All of this upset Washington greatly. And there were financial concerns. The only indication of the depth of the discouragement that they both felt is a letter of resignation, dated December 1875. It was never submitted. It may have been written by her without his knowledge, or it may have been dictated by him. We will never know.

But in July of 1876 the New York tower and anchorage were finished and he seemed to be better. The cable-spinning would soon begin, and that was the work he knew and liked best. There was political wrangling over whether the Roebling family should supply the wire, there was concern by the Trustees over his health, and by this time the unseen Chief Engineer had become an intriguing mystery to the public. For any and all of these reasons they moved to New York in October, and then back to Brooklyn in July of 1877. Once again he could watch the bridge growing from his own window, and now Emily's role was to become even more important.

At some point Emily began going to the bridge to "check on things for him." (McCullough 1972, p. 340) She took information with her from Roebling, and she returned with questions from the assistant engineers and her impressions of how things were going. One can only guess at the effect she had on the workers and the engineers when she first began her visits. Through it all she kept up with the great volume of written work generated by Roebling's concern for detail. Sometimes one letter would take an hour to dictate. In addition she kept his private day journals and a big scrapbook of all the newspaper articles about the bridge, but nothing about herself.

By 1879 she was 35 years old, and her son was 12 and in private school. By now her whole life was the bridge. The previous seven years had been somewhat like an intensive course in the design of a suspension bridge. "In truth she had by then a thorough grasp of the engineering involved. She had a quick and retentive mind, a natural gift for mathematics, and she had been a diligent student during the long years he had been incapacitated." (McCullough 1972, pp. 462–463) By this time she was visiting the bridge daily, sometimes two and three times a day.

In Trenton it had been relatively easy to avoid many visitors because of their distance from the bridge. Now, back in Brooklyn, she often found it necessary to talk, in her husband's place, to bridge officials, contractors' representatives, or visiting dignitaries. She did so with complete confidence, able to answer technical questions directly and to field political ones adroitly. "[M]any of them went away convinced she knew as much about the technical side of the bridge as any of the assistant engineers." (McCullough 1972, p. 463) Some of their return correspondence was addressed only to her.

Her role had progressed from wife to nurse to private secretary to administrative assistant, inspector and messenger. In the last stages of construction, Emily was his ceremonial representative, his ambassador, and his spokesman to the outside world. At this stage it was her ability to field political questions, to charm both sides in a controversy, and to play the part of peacemaker that were important.

By the summer of 1882 the difficult technical problems had all been dealt with. The Roeblings could take a small vacation, their first in five years. Emily rented a modest house in Newport, close to where her brother was stationed. (In fact, he died while they were there, not knowing that he had been cleared by the military court of inquiry, as he had hoped to be for so long.) But the Board of Trustees was restive. A decision to strengthen the bridge floor to make it capable of carrying railroad traffic caused controversy, and there was pressure to finish the bridge, which was getting behind schedule because of slow steel deliveries. The Board of Trustees wanted to confront the Chief Engineer with charges of "lack of leadership." Twice he was directed to attend a Board meeting, and twice he refused. Seth Low, mayor of Brooklyn, introduced a resolution to fire Roebling. At the same time a newspaper reporter printed an off-the-cuff remark of Roebling's that was unflattering to members of the Board. The situation looked extremely grave. Emily wrote a gracious letter of thanks to Ludwig Semler, a new member, when he spoke briefly in favor of Washington. That letter resulted in Semler's traveling all the way to Newport, and returning with a most favorable impression, which he then attempted to convey to the Board. Emily also wrote a letter of despair and apology to William Marshall, an older ally on the Board, just after the incident with the reporter. Marshall, who had never said much before, then spoke eloquently for Roebling during the all-crucial next Board meeting. The final board vote was against Roebling's resignation.

Emily appeared to enjoy the ceremonial duties. In 1881 she led a small procession for a "first walk" across the bridge on a plank walkway about five feet wide that had been built on the newly finished understructure for the bridge floor. She was toasted at various engineering gatherings. When Ferdinand de Lesseps visited New York to promote his planned Panama Canal project, Emily was one of the women who accompanied him on his grand entrance to the great banquet in his honor. But to be the first person to ride over the bridge in a carriage must have filled her heart with joy. She carried a live rooster as a symbol of victory. For the final great ceremony of the opening of the bridge, Emily arranged for a reception at their home to bring the President, the mayors, and the dignitaries all to Roebling's doorstep. His only ceremonial duty as Chief Engineer was to stand and receive the President of the United States in his own home.

Emily seems to have enjoyed the years following the end of the bridge construction very much, too. She worked on a book and took a law course at New York University. She drove her own carriages around the city. She built a new house and she entertained. She traveled to Europe, on one trip attending the coronation of the Tsar in Russia. But she became ill in December 1902, and died a few months later. She was almost fifty-nine years old. Roebling, incredibly, died in 1926 at the age of eighty-nine.

No will ever know how much Emily Roebling actually contributed to the bridge. She made a great effort to keep her role in the background and to maintain a high degree of privacy around their personal life. She gave no personal interviews and kept no memoirs. The records she did keep were only about the bridge and about Washington as Chief Engineer. I am sure that she would have felt her greatest tribute came from her husband when he called her his "strong tower," and referred to her as "a woman of infinite tact and wisest counsel." She was brought up to be a proper Victorian woman, always supportive of and secondary to her husband.

But Emily Roebling was a woman in a most unique situation for her time. She was a woman among men, able eventually to speak their language. That she was welcome among them, that her opinions were sought and respected, in an age when a woman's presence near a construction site was absolutely unheard of, is a great tribute to her talents and to what she had been able to learn.

I know that one must be careful about using the term "engineer" for someone who does not have the proper credentials. And I know that it is not clear that Emily ever did any actual engineering work on the bridge. But I feel that women engineers of today must look at her as an early pioneer, and early role model for themselves. Had she been alive today, the evaluation of her work would be different. She would have had a job title quite different from the only official one she ever had: wife.

Reference

McCullough, David. (1972). *The Great Bridge*, New York: Simon & Schuster.

7

1926 · *Kate Gleason's Careers*

Eve Chappell

KATE GLEASON *[1865–1933. ED.]*, OF ROCHESTER, NEW York, is a person with a triple career. As engineer, business woman and house constructor hers is an important name, and in each of these enterprises she has found life interesting and satisfactory. Also highly successful.

Miss Gleason was the first woman to qualify for full membership in the American Society of Mechanical Engineers, with all that means of practical experience. Secretary and treasurer of the famous Gleason Gear Planer Company for years, she had earlier done notable work for the foundry in selling its wares, and establishing agencies abroad. Her wanderings took her to the Orient, as well as Europe. During the war she was president of a national bank in East Rochester. For the last five years she has given her attention to house construction, bringing to bear on the building of small concrete houses the results of her years of experience and observation in the engineering field. Her aim is to produce small houses which shall be proof against both earthquake and fire, and being contracted for in large lots, can be sold at a much cheaper rate than now prevails.

But when a woman has three careers to her credit it is she herself that interests rather than any details of her work. Miss Gleason is gray-haired and plump; energetic, of course, humorous and very kindly. Somewhere in the middle fifties, one surmises; the genial, smooth-running fifties of

Chappell, Eve. (1926). "Kate Gleason's careers." *The Woman Citizen*, January, 19–20, 37–38.

Kate Gleason.
Courtesy of Jan Gleason, used with permission.

men and women to whom obstacles have been nothing but obstacles, after all, and not to be taken seriously. Her voice is full and low pitched. "I attended to that myself," she said. "Usually a woman's voice is too high, and the more energy she has, the higher it rises, and she wears everybody out. So I put mine where it is." That remark draws one telling line of her portrait. Her talk about clothes draws another. "Yes, I have lots of pretty dresses. That's one nice thing about having to go to Paris now and then. I think the pleasanter one can make oneself in whatever way the better for everybody and everything concerned." After that one need not dread that the concrete houses will be ugly. Cheap and indestructible they will be, since that is what Miss Gleason is trying for, but they will surely have about them some aspect of charm.

Very early Kate Gleason began to take stock of life; to measure conditions and to meet them, not with surrender, but with intelligent circumvention. "Girls were not considered as valuable as boys; so I always jumped from a little higher barn, and vaulted a taller fence than did my boy playmates, just to prove that I was as good as they." When she was eleven, her twenty-year-old stepbrother died, and she heard her engineer father lament that in the loss of his son he had also lost a helper; some one to go on with his work. Then and there Kate Gleason made her decision about a career; she told her father at once that she was going to be an engineer. Three years later she began her training and work with him. That this was permitted was little short of being a town scandal. Nothing but stinginess could account for a girl being allowed to do such work. But Kate Gleason's father was a friend of Susan B. Anthony, and was willing to abide by her doctrines even when they struck home as closely as that. So Kate worked in the shop, laboring with tables and designs and more technical intricacies than she had bargained for, but not more than she was eager to master. This training with her father was supplemented by a course at Cornell College, the only girl engineer in the class of '89.

Has my work been made harder because I am a woman? No. I have no hard knocks to report. Indeed, I think engineering must be different from any other profession in that regard. Engineers are as a class so successful, so progressive that they bear no grudges; feel no jealousies for any newcomers into their ranks, whether man or woman. And besides, I had the advantage of being my father's daughter, and he ranked high. Impossible to estimate the help this may have been to me. Associations count for much in any success and unfortunately they are in great degree a matter of chance. Still, when I recall stories told me by women struggling for place in other professions I insist that engineers are in a class apart.

When Miss Gleason was nineteen she first went on the road to sell certain tools or appliances designed by her father; in wholesome fear of seeming to advertise, she is not more explicit. This was another departure from custom; another test of his willingness to put his principles to the proof. A small concession he made to prejudice; advance letters indicated that his daughter would be in the town to see an exhibit, and the sales errand was made to appear an unimportant side issue. Blue-eyed, black-lashed Kate Gleason allowed the subterfuge to stand. It was taking orders that mattered, and she always got the orders. Several years later she made her first business journey to Europe, one of the first made by any American manufacturer. In demonstrating her tools and gears, her engineering knowledge was invaluable. Not in her sales work either was she hindered by being a woman. Gaily she said:

> That 1894 European trip took two months, and cost two hundred dollars from Rochester to Rochester. I went on a cattle steamer, the Mongolian, from Montreal, and I still consider it a most luxurious and pleasant trip. I was the only woman passenger and my partners for promenades walked with me two at a time, under stop watches held by the next in line. The purser told me of a trip he had made into the Congo, when for six months he did not see a white woman. On the way home a colonel's wife came on board, and she seemed to the purser so beautiful that it was rapture merely to stand and gaze at her. But at Madeira they picked up a number of fresh young English girls, and to the purser's amazement, he at once noticed that the colonel's wife had a leathery complexion and false teeth. I saw the point of the purser's story, and ever since I have been developing a talent that almost amounts to genius for putting myself in places where other women are not likely to come.

In support of this claim, it may be said that she was the only woman among eight hundred men on the main floor at the dinner of the American Society of Mechanical Engineers held in New York early in December. Also, she is the only woman at the meetings of the American Concrete Institute.

With the years the Gleason works have grown and prospered. So Miss Gleason has taken up the house-building enterprise. "It's a whole new life." Miss Gleason says. "Starting a new business in middle age: new problems, new conditions to cope with—it's splendid. I have observed the methods of the automotive engineers, who are miracle workers, and by putting what I could of these methods into house building I am making a success of it."

In her building Miss Gleason is working at one of the real problems facing this country: that of adequate, substantial and economical housing for the families of workers. She has tackled the problem as an engineer, and has made valuable contributions to building practice. It began by her taking over an unfinished building project, an obligation of the bank of which she was president during the war years and considered by that bank a doubtful liability. The project called for the building, at lowest possible rates, of standard frame houses of given dimensions; the sort of which one must make careful count from the corner lest one fit one's latch-key into the wrong door. Miss Gleason believed that houses could be built which would be convenient, pleasing, differentiated, and yet inexpensive. During 1919 and 1920 she worked out her plan for the building of one hundred six-room dwellings; fireproof, substantial enough to stand for centuries, and to cost four thousand dollars. They are built of concrete; have two stories and a basement given over to garage, furnace and laundry, and are so arranged that the similarity of design is concealed.

Modifications have been made in Miss Gleason's later houses, and she has experimented as no contractor may. Two years more she plans to give to experimentation. In California she will study the work in stucco, coloring and house placement with regard to lay of lots and surrounding scenery. Further study is to be made in France. Then Miss Gleason will take up house construction on a large scale. Meanwhile the work of her present organization goes on at the rate of sixteen houses a year.

Miss Gleason has lately turned over to the care of a trust company as much as possible of her other interests. She gives two reasons. "I want to guard against making queer investments and unwise changes in my will, as women and other people are so likely to do as they get on in life." But her second reason is stronger; she wishes to give herself whole-heartedly to the new enterprise, and not be distracted by other affairs. This philosophy of whole-heartedness—one thing at a time so far as may be—is responsible for the fact that she is not married. "Marriage is a career all by itself," she says. "Some women do not find it so. But it would have had to be with me."

Miss Gleason is exuberantly well. She can walk her thirty miles a day easily. "I find my pleasure in concrete things rather than in people," she said, and added a bit of smiling life philosophy, which showed that she brings her engineering mind to bear on her personal relations. "I don't waste any time on antagonism, trying to make people love me who don't, or trying to make them live their lives the way I think they should. Too much friction is bad. I don't waste my energy that way. And I've had a wonderful life."

8

1993 · *Ahead of Their Time*

A Century Ago, Women Engineers—Such as the Brilliant Bertha Lamme— Blazed a Lonely Trail.

Jeff Meade

THE STORY IS TOLD OF A SALES ENGINEER FOR THE WEST-
inghouse Electric Works back in the 1890s who returned
from his travels with a problem. A client was demanding
major modifications to a motor. The sales rep called the lab
and asked to speak with the supervisor. The woman who
answered said, "He's not here. Perhaps I can help you."
With that, she started asking questions about the motor,
getting more and more technical, eliciting ever more star-
tled responses from the sales engineer. "Suddenly, he realized that the woman was redesigning the motor just about
the way the customer wanted," says Westinghouse historian
Charles Ruch. "After he hung up the phone, the salesman
ran over to the lab to see what kind of secretary could do
that kind of thing."

She was no secretary but one of the first women engineering graduates in the nation, Bertha Lamme *[1869–1943. Ed.]*.

Meade, Jeff. (1993). "Ahead of their time." *ASEE Prism*, January,
27–29. Reprinted with permission of the American Society for Engineering Education.

Bertha Lamme.
Courtesy of IEEE History Center.

In 1893, the year ASEE *[American Society for Engineering Education]* was founded, Lamme received her bachelor's degree in mechanical engineering from Ohio State University, with a special emphasis in electricity. (At the time, there were about 200 programs nationwide granting similar degrees.) Bertha was the younger sister of Benjamin Carver Lamme, an 1888 Ohio State engineering graduate, who was already making a name for himself as chief engineer for George Westinghouse's company. Later he would design the first Niagara Falls generators. Like her brother before her, Bertha went to work for Westinghouse. She designed and built motors.

"Being Benjamin's sister probably helped pave the way for Bertha," says Ruch. "But she was brilliant in her own right. After she graduated from Ohio State, she apparently impressed Albert Schmid, who was then a shop superintendent at Westinghouse. He is the one, more than her brother, who was responsible for hiring her."

Though Bertha Lamme had a brief, illustrious career in industry, there are indications that she might have pursued an engineering degree simply out of intellectual curiosity. In his autobiography, Benjamin Lamme writes: "She had taken an engineering degree at Ohio State University, more for the pleasure of it than anything else; but sometime later, Mr. Schmid gave her a serious invitation to enter the employ of the company, so she took up the work of calculation of machines and stayed until she married."

Whether or not she set out to make a name for herself, Bertha Lamme succeeded in doing so. A 1907 *Pittsburgh Dispatch* account of her tenure at Westinghouse said that Lamme's work in designing dynamos and motors won her a reputation, "even in that hothouse of gifted electricians and inventors. She is accounted a master of the slide rule and can untangle the most intricate problems in ohms and amperes as easily and quickly as any expert man in the shop."

Reputation or no, Bertha Lamme was obliged to leave Westinghouse in 1905 when she married Russell S. Feicht, director of engineering. Westinghouse, along with just about every other major corporation in America at the time, had a policy requiring women employees to quit after they married another company employee. Though brothers and sisters could work together, wives were not permitted to remain on the same payroll with their husbands, says Ruch.

Bertha Lamme was a trailblazer, on a lonely trail indeed, but she had some company. According to a history of the Society of Women Engineers compiled by SWE past president Arminta Harness, at least one woman had earned an engineering degree before 1893. In 1876, a full 17 years before Bertha Lamme received her degree, the University of California at Berkeley graduated a young woman named Elizabeth Bragg. She took a degree in civil engineering. Harness thinks that Bragg may be the first degreed woman engineer in the country. The first to practice engineering may have been Edith Julia Griswold, who began studying both law and engineering in 1884, and just two years later opened an office in New York City specializing in patent office drawings. She was the first woman listed in *Who's Who of Engineering*.

Other evidence suggests that Bertha Lamme was not alone. The U.S. Census of 1890 counted 21 women as engineers. Whether any of them had degrees is unknown. (The SWE history mentions an informal survey of engineering schools that yielded a total of 16 engineering degrees granted to women before 1919.) Presumably, many who did engineering work lacked formal training. One often cited is Emily Roebling, now believed to have finished work on the Brooklyn Bridge after her husband Washington became ill.

If Bertha Lamme commands more attention than other women engineers of the time, it is partly because we have more of her personal history. Others may be overlooked because no one thought to keep records of them. It is also possible says Harness, who is SWE's unofficial historian, that many of them never worked as engineers. "The same thing probably happened to them that happened to women engineering graduates in the 1950s," she says. "Even if they did work for a brief period of time they got married and quit."

It is no secret that in the late nineteenth century, women who persisted in practicing engineering met with little encouragement from male colleagues. Historical records show, however, that some engineers did keep an open mind on women in the profession. For example, Mrs. Roebling's accomplishments attracted the attention of DeVolson Wood, president in 1894 of the fledgling Society for the Promotion of Engineering Education *[SPEE]*. In his address at the 1894 annual meeting in Brooklyn, Wood said: "Who would not feel that this Society would be honored by enrolling among its members that woman who, when her husband's health was being undermined, studied works on engineering and the plans and specifications of the structure, and became the head, hands, and feet of him who was the official engineer of the East River bridge, Roebling?"

Wood, obviously a man ahead of his time, foresaw the day when women would take their place alongside men in engineering, as well as engineering education. Indeed, the constitution of the SPEE, forerunner of ASEE, did not specify whether the society should be closed to females. Commenting on this, Wood said "[I]n these days of great enlargement of the educational field for women, it looks like a stroke of wisdom to so frame the organic law that the admission of women to the Society might be possible without calling a constitutional convention and entering into political strife over such a minor point."

From Computor to Electrical Engineer

The Remarkable Career of Edith Clarke

James E. Brittain

Edith Clarke's electrical engineering career had as a central theme the development and dissemination of mathematical methods that served to simplify and reduce the time spent on laborious calculations in solving problems encountered in the design and operation of large electrical power systems. As an engineer with the General Electric Company from the early 1920's to 1945, she worked during a time when power system analysis was evolving from being labor intensive to being machine intensive, with much of the labor of problem solving being shifted from human computors, often women, to electromechanical computers, such as the network analyzer and differential analyzer. *[As in the original, "computor" refers to human beings while "computer" refers to machines. Ed.]* This trend culminated in the development of electronic computers beginning with the

Brittain, James. (1985). "From computor to electrical engineer: The remarkable career of Edith Clarke." *IEEE Transactions on Education*, E-28(4), November, 184–189. © IEEE. Reprinted with permission of IEEE and the author.

141

Edith Clarke.
SWE Archives, Walter P. Reuther Library, Wayne State University, used with permission.

ENIAC that was completed during the same year that she retired from GE. *[For more on the ENIAC, see Chapter 11. Ed.]* As a woman who worked in an environment traditionally dominated by men, she demonstrated that women could perform engineering analysis at least as well as men if given the opportunity. Her achievements provided an inspiring example for the next generation of women with aspirations to seek a career in electrical engineering.

Introduction

Edith Clarke (1883–1959) began her career as a Computer Assistant to a Research Engineer at the American Telephone and Telegraph Company. Eventually, she overcame formidable entry barriers to women and became a professional Electrical Engineer. She specialized in the analysis of electrical power systems and was employed by the General Electric Company from the early 1920's until her retirement in 1945. She then became a Professor at the University of Texas where she taught until her second retirement in 1956. Clarke was the first woman to receive a degree in electrical engineering from the Massachusetts Institute of Technology (MIT), Cambridge, and the first woman to present a technical paper before the American Institute of Electrical Engineers (AIEE).

Clarke directed much of her work as an engineer toward the simplification and mechanization of laborious calculations encountered in the analysis of transmission lines and power systems. Somewhat ironically, her most important contributions to engineering analysis tended to reduce or eliminate the need for skilled human computors, an occupation that had come to be regarded as suitable work for women at a time when engineering was not. During the time spanned by Clarke's working career, power systems analysis evolved from being labor intensive to being machine intensive, with much of the labor of problem solving being shifted from human computors, often women, to electromechanical computers. This trend culminated in the development of electronic computers beginning with the ENIAC that was introduced during the same year that she retired from General Electric.

Clarke helped to develop and teach mathematical models that took advantage of such electromechanical aids as calculating tables and alternating current network analyzers. In effect, she wrote what now would be called software for the machines that set the stage for electronic digital computers. She became highly skilled in the manipulation of hyperbolic functions and symmetrical components, and was able frequently to simplify their use by preparing graphs and tables. Early in her career, she invented and patented a graphical calculator for the solution of transmission line problems. Her contributions made the "woman's work" of unmechanized computing less necessary and, in the process, initiated the arduous task of opening the traditionally male-dominated electrical engineering culture to women.

Early Life and Education

Edith Clarke was born on February 10, 1883 and spent her early years on a farm in Maryland near Ellicott City. She was one of nine children of John R. Clarke, a lawyer-farmer, and Susan Owings Clarke. In her early childhood, Edith Clarke suffered from what probably now would be diagnosed as a "learning disability" in reading and spelling, but she exhibited a good aptitude for mathematics and enjoyed puzzles and card games, especially duplicate whist. Both her parents died by the time she was 12 and an uncle served as her legal guardian. She attended a boarding school in Montgomery County, MD, until 1899. At the age of 18, she received a modest inheritance from the estate of her parents and decided to use it to continue her education. She enrolled at Vassar College in Poughkeepsie, NY, a college that had opened in 1865 and that provided "the real impetus toward the full collegiate education of women".[1] At Vassar, she concentrated on mathematics and astronomy, areas where the college already had established a strong reputation. Many Vassar alumnae in the late 19th and early 20th centuries found employment at observatories as computational assistants to male astronomers.[2]

After receiving an A.B. degree from Vassar in 1908, Clarke taught mathematics and physics for a year at a school for girls in San Francisco, CA. She then taught mathematics for two years at Marshall College in Huntington, WV. She expressed her discontent at the prospects of a career as a mathematics teacher by enrolling as a sophomore in civil engineering at the University of Wisconsin in the fall of 1911 at age 28.[3]

A Computor at AT&T and a Student at MIT

After spending an enjoyable year as an undergraduate engineering student, Clarke was hired for the summer as a Computer Assistant to George A. Campbell (1870–1954), an outstanding if somewhat reclusive Research Engineer with AT&T. Clarke found the computing work sufficiently interesting so that she abandoned her plan to return to Wisconsin to complete the requirements for an engineering degree. Campbell had been a Bell employee since 1897 and already had played a major role in the development of the loading coil, a major innovation in telecommunication.[4] At the time Clarke was assigned to work for Campbell, his efforts were directed to the analysis of problems related to the use of vacuum-tube amplifiers on long distance telephone lines. The Bell Company had embarked on a crash program to complete a transcontinental line from New York to California, to be operational in time for a planned celebration of the completion of the Panama Canal. Consequently, the telephone company was expanding its research effort on related innovations such as repeater amplifiers. Thus, the job opportunity for Clarke was in the context of this accelerated research program that provided work for several women with credentials in mathematics.[5]

Clarke's time as assistant to Campbell provided her with an excellent apprenticeship in the mathematical theory of transmission lines and electric circuits. Campbell was the company's leading authority on these topics that involved the manipulation of hyperbolic functions, equivalent circuits, and graphical analysis, areas that Clarke pursued for the rest of her career. Among the computational tasks assigned to her was to calculate the first seven terms of an infinite series that represented a probability function. Another woman computor, Sallie E. Pero, extended the series to eleven terms while Lucy Whitaker used a different method to provide an independent check of the work done by Clarke and Pero.[6] In 1915, Clarke joined the Transmission and Protection Engineering Department at AT&T where she was responsible for training and directing a small group of computors. During this period, she also found time to take a course in radio at Hunter College and several night classes at Columbia, New York, NY.

In 1918, Clarke took a decisive step toward becoming an engineer when she left AT&T to enroll in electrical engineering at MIT. She completed the senior level undergraduate courses by the fall of 1918 and continued on to receive the Master of Science degree in electrical engineering in 1919. Although she was not the first woman to graduate from MIT, she was the first woman to earn a degree in electrical engineering from the school. Arthur E. Kennelly and Vannevar Bush were leading members of the MIT faculty while Clarke was a student there. Kennelly taught transmission line and advanced alternating current circuit theory. He was known for his ability to simplify difficult concepts and mathematical methods and express them in terms that the average engineer could understand, a talent that Clarke also displayed as an engineer.[7] Bush came to MIT as a Professor in 1919 and achieved a reputation for his skill in electrical analysis by means of operational calculus and for developing the network analyzer and the differential analyzer, electromechanical calculators of the type exploited by Clarke and her colleagues during the 1930's.[8]

A Computor at GE and an Invention

Despite her newly acquired graduate degree, Clarke experienced difficulty finding employment as an engineer. Again she became a computor, this time with the Turbine Engineering Department of the General Electric Company in Schenectady, NY. During the period from 1919 to 1921, she trained and directed a small team of women computors in the calculation of mechanical stresses in high speed turbine rotors. As in her earlier computing job at AT&T, the position with GE resulted from an anomalous situation that created a temporary need for skilled calculators. The increasingly large turboalternator units manufactured by GE just prior to 1920 developed unanticipated problems caused by rotor vibrations and metal fatigue. In response, GE launched an intensive research effort directed by C. E. Eveleth that involved the efforts of several engineers and members of the GE Research

Laboratory. The group carried out a program that included both experiments and theoretical analysis. The GE researchers developed a comprehensive theory of rotor vibrations and developed techniques to overcome the problem at normal operating speeds. They disseminated their findings in a number of technical papers published during 1924–1925.[9]

In June 1921, Clarke filed a patent application describing her invention of a graphical calculator to be used in the solution of transmission line problems. The invention probably was based on her earlier work at AT&T or as a student at MIT. Since she was not a salaried engineer, she was not required to assign the patent rights to GE. The patent description included instructions on how to construct and assemble the device. She explained that the purpose of the calculator was to facilitate the analysis of transmission lines of different lengths or different characteristics by reducing the need for laborious computation. She revealed that the graphical calculator was based on line equations that included hyperbolic functions, and she cited Oliver Heaviside's book, *Electromagnetic Theory*, and A. E. Kennelly's *The Application of Hyperbolic Functions to Electrical Engineering Problems*. The patent was issued in September 1925.[10]

Clarke's graphical calculator was the subject of her first technical paper published in the *GE Review* in 1923.[11] The paper contained drawings of the parts of the calculator along with instructions on how to mount the parts on cardboard and assemble. The device included a base chart and two calibrated radial arms that were attached to the base chart. She stated that her calculator served to reduce "time and labor" by providing a graphical solution of line equations that took into account distributed resistance, inductance, and capacitance. She explained that the calculator gave quite accurate results for power transmission lines of up to 250 miles long, and that a problem solution required less than a tenth of the time ordinarily required. Her paper included a derivation of equations relating line current, voltage, impedance, and admittance that could be solved by means of her calculator. The function of the Clarke calculator was similar to the well-known "Smith chart" developed by P. H. Smith of the Bell Telephone Laboratories in the late 1930's.

In 1921, Clarke left GE to accept a one-year appointment teaching physics at a woman's college in Istanbul, Turkey. She utilized the opportunity to tour several European countries and Egypt before returning to Schenectady in 1922 to sign a contract as a salaried electrical engineer at GE. At age 39, she finally had achieved her goal of acceptance as an engineer instead of a computer assistant to engineers. She soon took advantage of the broader horizon open to an engineer by becoming an active member of the AIEE and a licensed Professional Engineer in New York. At GE, she joined the Central Station Engineering Department.[12] She used her early work in telephone transmission theory to great advantage and helped to educate power engineers in analytic methods from telecommunication engineering that increasingly were needed as power lines increased in length to appreciable fractions of a wavelength.

A GE Engineer

In April 1925, Clarke applied for a patent on a method designed to regulate the voltage on power transmission lines. The patent was issued in September 1927 and was assigned to GE. The invention utilized a rotary machine with special saturable field poles that acted as a synchronous phase modifier. Clarke explained that the phase modifier was intended to add sufficient reactive power to the system to prevent an excessive drop in the terminal voltage. This enabled the transmission line to be operated nearer its maximum power limit, an important consideration in long high voltage lines.[13]

Clarke was the author or co-author of 18 technical papers published between 1923 and 1945. In February 1926, she became the first woman to present an AIEE paper. The paper, later published in the *Transactions of the AIEE,* was entitled "Steady-state stability in transmission systems-calculation by means of equivalent circuits or circle diagrams".[14] She pointed out that the growing trend toward longer lines and greater system loading had made it necessary to give more attention to the problem of system stability. She explained how an actual power system could be analyzed by means of an equivalent circuit with lumped-constant elements. As a source, she cited A. E. Kennelly who had included a chapter on equivalent circuits in his book *Hyperbolic Functions Applied to Electrical Engineering.* Clarke then introduced a correction factor containing hyperbolic functions that she used to produce a more exact equivalence with the actual system. The modified equivalent circuit then was used to calculate the maximum power that could be carried without instability.

During the discussion of Clarke's paper, Charles L. Fortescue of the Westinghouse Electric Company expressed his appreciation for the "ingenious method" that she had developed. He also complimented the "able way" in which she had presented material on such a difficult subject. Robert D. Evans, also a Westinghouse engineer, remarked that the paper was of "very considerable interest" and he anticipated that the equivalent circuit method would "find general use." H. H. Spencer commented that the profession should be grateful to Clarke for having simplified such a difficult problem to a "slide rule and arithmetical basis".[15] Fortescue and Evans, along with

Clarke, played a major role in the development of the method of symmetrical components for power system analysis.

In May 1926, Clarke published a tutorial paper in the GE *Review* entitled "Simplified transmission line calculations".[16] In an introductory note, the editor observed that few engineers had mastered some of the more "specialized branches of higher mathematics." He continued that Clarke had developed a simpler approach for the use of hyperbolic functions in problems frequently encountered. In the paper, Clarke also noted that "hyperbolic functions, real or complex, are not popular with engineers," and that the equivalent series had met with "greater favor." However, she pointed out that the labor required increased with the number of terms in the series and became very great in the case of long lines. Consequently, she had developed graphs for hyperbolic functions of the form often used in engineering analysis. She included examples of how to use the graphs to obtain quite accurate results.

During the 1920's, Clarke learned to use the method of symmetrical components in the analysis of polyphase power systems. This powerful method essentially did for the analysis of the multiple-phase power system what the method of complex numbers had done for the analysis of single-phase systems when it was introduced in the 1890's by A. E. Kennelly and Charles P. Steinmetz. The symmetrical components method facilitated analysis of unbalanced three-phase circuits by converting the problem into equivalent symmetrical circuits that were much simpler to analyze. Clarke credited L. G. Stokvis as having been the first to publish on a method to resolve an unbalanced electrical system into positive and negative sequence components. He published a paper on the method in a German periodical in 1912 and a second paper in the *Electrical World* in 1915. In the 1915 paper, Stokvis actually used the terms synchronous components and inverse components for what later were called positive and negative sequence components.[17]

C. L. Fortescue presented a more general analysis that included a zero sequence component in addition to the components used by Stokvis in a classic AIEE paper published in 1918.[18] Fortescue explained that any system of three vectors (a term then commonly used for complex quantities or phasors) could be expressed by three sets of balanced components that he called symmetrical coordinates. During the discussion of Fortescue's paper, Vladimer Karapetoff suggested that symmetrical components was a "more correct and descriptive term" for the method than symmetrical coordinates. The method was disseminated and extended to practical problems in the field of power systems in a number of papers published during the 1920's. In 1925, R. D. Evans published a paper in *Electrical World* on the use of symmetrical components in the determination of short-circuit currents.[19] He also explained how the method could be used in conjunction with the so-called "calculating board" that was used to simulate actual power systems. Calculating boards or tables had been used at least since 1916 to determine short-circuit currents in alternating-current systems by simulation using banks of rheostats to represent actual generators, lines, and loads.[20] Also published in 1925 was a paper by Sadatoshi Bekku, a Japanese engineer, that discussed the use of symmetrical components and the calculating table to determine short-circuit currents in three-phase power systems.[21] Charles F. Wagner and R. D. Evans subsequently published a tutorial series on symmetrical components in the *Electric Journal* during the period from March 1928 to November 1931.

Clarke presented her first paper using the method of symmetrical components at an AIEE meeting in March 1931.[22] She extended the method that others had used to determine the effect of single faults to the analysis of two or more simultaneous faults involving a three-phase transmission system. She mentioned that a calculating table or the MIT network analyzer could be used to facilitate the analysis. During the discussion of Clarke's paper, an engineer employed by a California utility commented that the paper had "great practical value" and represented "an important step forward in protection engineering." Harold W. Bibber credited Clarke with pointing the way to the solution of "almost any problem involving simultaneous faults that might arise." Clarke and Bibber began work on a book on symmetrical components the following year but did not complete the project although portions of the manuscript eventually were incorporated in Clarke's book published in 1943. Harold L. Hazen, a Professor at MIT who had been in charge of the network analyzer project that had been supported financially by GE, was in the audience when Clarke's paper was presented and gave a brief commentary on the machine and its use in power system analysis. The MIT network analyzer was regarded as "the most advanced system for transmission network analysis in the world" in the early 1930's, and approximately 40 of the machines were built for use by utilities, electrical manufacturers, and engineering schools.[23]

Clarke's third AIEE paper presented in May 1932 was awarded a prize as the best paper of the year in the northern district of the AIEE.[24] The subject was multiple conductor transmission lines where two or more conductors were used per phase in a three-phase power line. She mentioned that there had been a revival of interest in such systems and their potential to increase line capacity. She included equations and graphs for lines having from two to five conductors per phase in various geometrical arrangements such as three configured as an equilateral triangle, four as a square, or five in a vertical collinear array. In her analysis, she compared multiple conductors to single conductors with regard to the charging current, no-load voltage, and power capacity. She again employed symmetrical components in the analysis.

A differential analyzer at the Moore School of Electrical Engineering of the University of Pennsylvania was used to collect data for an AIEE paper by Clarke and two co-authors presented in January 1938.[25] Her co-authors were Cornelius N. Weygandt, an instructor at the Moore School, and Charles Concordia, a GE colleague. The differential analyzer at the Moore School was patterned after the machine developed by Vannevar Bush, Harold Hazen, and associates at MIT, by 1930, and it set the stage for the famous ENIAC, the electronic digital computer developed at the Moore School during World War II. The Moore School differential analyzer had been undertaken as an emergency relief project with funding from the Civil Works Administration and Federal Emergency Relief Administration and had employed 120 men at the peak during its construction. The designers were encouraged to maximize the "ratio of labor cost to material cost" with the result that many parts of the machine were designed to be made by machinists assigned to the project that might have been purchased more cheaply. Somewhat ironically, the ultimate purpose of the machine itself was to reduce greatly the labor of engineers and human computors.[26] The paper by Clarke and her co-authors dealt with the problem of voltage surges produced by unbalanced faults and how overvoltages could be reduced by means of so-called "amortisseur windings" or damper windings on the generators supplying the fault current. The authors mentioned that the use of the differential analyzer had enabled them to calculate a wide variation in the key parameters that would not have been feasible in field tests or without the aid of the calculating machine.

Clarke contributed significantly to the development of the method of modified symmetrical components. She published a tutorial paper in the *GE Review* in 1938 on the subject of alpha, beta, and zero components that were derived from the conventional symmetrical components.[27] She included several examples of problems that were simplified by the use of the modified components. She later credited Walter W. Lewis with having used the modified components without the names in a paper published in 1917. The modified components were also sometimes known as x, y, and z components. In her book published in 1943, Clarke observed that the definitions and notation for the modified components still were "not definitely established by usage." She devoted a full chapter of her book to the use of alpha, beta, and zero components in the analysis of three-phase systems.[28]

In January 1941, Clarke and Selden B. Crary, a GE engineer, presented a joint paper at an AIEE meeting in Philadelphia, PA, that was awarded a prize as the best AIEE paper of the year.[29] Their paper contained a comprehensive analysis of the problem of stability on long transmission lines with much of the data being obtained using a network analyzer. They considered lines with lengths of up to a full wave length at 60 Hz, and discussed eight different methods that had been proposed to enhance stability of long lines. Their analysis indicated that the use of series capacitance was the most effective method of maintaining stability on long alternating current lines of up to 700 miles in length.

In 1943, Clarke published the first of a planned two volume work on the circuit analysis of alternating-current power systems. The book was based on her notes for lectures given over a period of many years to engineers of the Central Station Engineering Department of GE and was intended for use as a text by engineering schools or as a reference for power system engineers. The principal focus of the book was on the use of symmetrical components and modified symmetrical components in the solution of problems related to polyphase power transmission. In the introduction to the book, she wrote that the problems encountered by the power transmission engineer at any given time could be divided into three classes. Problems of the first class could be "solved analytically by well known methods in general use" in a reasonable amount of time. She continued that problems of the second class "can be solved analytically and the various factors evaluated, but the time and labor required are excessive." For problems of the third class, she noted that "there is no known analytic method of evaluating all the factors involved" and, therefore, "a different and independent problem is encountered with each change in given conditions." She observed that the calculating table, the network analyzer, and the methods of symmetrical components had helped move many problems from class two to class one and some from class three to class two. She pointed out that many problems that involved nonlinear parameters remained in the third class, although a few such problems had been solved by means of the differential analyzer or a transient analyzer.[30]

A Professor of Electrical Engineering

Clarke retired from GE on July 31, 1945 but accepted an opportunity to teach electrical engineering at the University of Texas in the spring of 1947. Her initial intention was to teach for only four months, but she returned to teach in the fall and continued to teach at the university until 1956. In 1948, she became the first woman to be made a Fellow of the AIEE. The second volume of her work on power system analysis was published in 1950. She was honored by the Society of Women Engineers in 1954 for her significant contributions to engineering. She also became the first alumna member of the Tau Beta Pi and Eta Kappa Nu scholastic honorary societies. She again retired at age 73 in 1956 and returned to her native Maryland where she died on October 29, 1959.[31]

Edith Clarke's engineering career had as its central theme the development and dissemination of mathematical methods that tended to simplify and reduce the time spent in laborious calculations in solving problems in the design and operation of electrical power systems. She translated what many engineers found to be esoteric mathematical methods into graphs or simpler forms during a time when power systems were becoming more complex and when the initial efforts were being made to develop electromechanical aids to problem solving. As a woman who worked in an environment traditionally dominated by men, she demonstrated effectively that women could perform at least as well as men if given the opportunity. Her outstanding achievements provided an inspiring example for the next generation of women with aspirations to become career engineers.

Notes

1. M. W. Rossiter, *Women Scientists in America*. Baltimore, MD: The Johns Hopkins University Press, 1982, p. 9.
2. M. W. Rossiter, "'Women's work' in science, 1880–1910," *ISIS*, vol. 71, pp. 381–387, 1980.
3. A. C. Goff, *Women can be engineers*, Youngstown, OH, 1946, Edith Clarke File, General Electric Company, Schenectady, NY. I am indebted to George Wise for providing me with a copy of this and other documents from the Clarke File.
4. J. E. Brittain, "The introduction of the loading coil: George A. Campbell and Michael I. Pupin," *Technol. Culture*, vol. 11, pp. 36–57, 1970.
5. L. Hoddeson, "The emergence of basic research in the Bell System, 1875–1915," *Technol. Culture*, vol. 22, pp. 512–544, 1981. Also see, *The Collected Papers of George Ashley Campbell*. New York: The American Telephone and Telegraph Company, 1937, p. 533.
6. *The Collected Papers of George Ashley Campbell*. New York: The American Telephone and Telegraph Company, 1937, p. 236.
7. J. E. Brittain, "Kennelly uses complex quantities to simplify AC analysis," *Proc. IEEE*, vol. 72, p. 462, 1984.
8. T. P. Hughes, *Networks of Power*. Baltimore, MD: The Johns Hopkins University Press, 1983, pp. 376–377. Also see S. Bennett, "Harold Hazen and the theory and design of servomechanisms," Univ. Sheffield, England, Res. Rep. 270, Nov. 1984.

9. P. L. Alger, *The Human Side of Engineering*. Schenectady, NY: Mohawk Development Service, Inc., 1972, pp. 71–72. W. Campbell, "The protection of steam-turbine disc wheels from axial vibration," *Trans. ASME*, vol. 46, pp. 31–140, 1924. W. Campbell, "Tangential vibration of steam turbine buckets," *Trans. ASME*, vol. 47, pp. 643–654, 1925. A. L. Kimball, Jr., and E. H, Hull, "Variation phenomena of a loaded unbalanced shaft while passing through its critical speed," *Trans. ASME*, vol. 47, pp. 673–688, 1925.

10. "Calculator," U.S. Patent 1 552 113, Sept. 1, 1925.

11. E. Clarke, "A transmission line calculator," *General Electric Rev.*, vol. 26, pp. 380–390, 1923.

12. Goff, *Women can be engineers*, Edith Clarke File, pp. 55–57.

13. "Electrical power transmission," U.S. Patent 1 641 737, Sept. 6, 1927.

14. E. Clarke, "Steady-state stability in transmission systems-calculation by means of equivalent circuits or circle diagrams," *Trans. AIEE*, vol. 45, pp. 22–41, 1926.

15. ———, "Steady-state stability in transmission systems-calculation by means of equivalent circuits or circle diagrams," *Trans. AIEE*, vol. 45, pp. 80–86, 1926.

16. ———, "Simplified transmission line calculations," *General Electric Rev.*, vol. 29, pp. 321–329, 1926.

17. L. G. Stokvis, "Analysis of unbalanced three-phase systems," *Electrical World*, vol. 65, pp. 1111–1115, 1915. Also see, E. Clarke, *Circuit Analysis of A-C Power Systems*. New York: Wiley, 1943, p. 54.

18. C. L. Fortescue, "Method of symmetrical co-ordinates applied to the solution of polyphase networks," *Trans. AIEE*, vol. 37, pp. 1027–1115, 1918.

19. R. D. Evans, "Finding single-phase short-circuit currents on calculating boards," *Electrical World*, vol. 85, pp. 760–765, 1925.

20. W. W. Lewis, "Calculation of short-circuit currents in alternating-current systems", *General Electric Rev.*, vol. 22, pp. 140–145, 1919. W. W. Lewis, "A new short circuit calculating table," *General Electric Rev.*, vol. 23, pp. 669–671, 1920.

21. S. Bekku, "Calculation of short-circuit ground currents on three-phase power networks using the method of symmetrical co-ordinates," *General Electric Rev.*, vol. 28, pp. 472–478, 1925.

22. E. Clarke, "Simultaneous faults on three-phase systems," *Trans. AIEE*, vol. 50, pp. 919–939, 1931.

23. S. Bennett, "Harold Hazen and the theory and design of servo- mechanisms," Univ. Sheffield, England, Res. Rep. 270, Nov. 1984, pp. 6 and 12.

24. E. Clarke, "Three-phase multiple-conductor circuits," *Trans. AIEE*, vol. 51, pp. 809–821, 1932.

25. E. Clarke, C. N. Weygandt, and C. Concordia, "Overvoltages caused by unbalanced short circuits—Effect of amortisseur windings," *Elec. Eng.*, vol. 57, pp. 453–466, 1938.

26. I. Travis, "Differential analyzer eliminates brain fag," *Mach. Design*, vol. 7, pp. 15–18, July 1935. See also, J. Brainerd, "Genesis of the ENIAC," *Technol. Culture*, vol. 17, pp. 482–488, 1976.

27. E. Clarke, "Problems solved by modified symmetrical components," *General Electric Rev.*, vol. 41, pp. 488–494 and 545–549, 1938.

28. ———, *Circuit Analysis of A-C Power Systems.* New York: Wiley, 1943, pp. 310–311.

29. E. Clarke and S. B. Crary, "Stability limitations of long-distance A-C power transmission systems," *Elec. Eng.*, vol. 60, pp. 1051–1059, 1941.

30. E. Clarke, *Circuit Analysis of A-C Power Systems.* New York: Wiley, 1943, preface, pp. 1–2.

31. "Edith Clarke, noted engineer and author retires from G.E.," *General Electric General Office News*, Aug. 10, 1945. "Teaching opens new world to woman, 65," *The Dallas Morning News*, Dec. 12, 1948. "Miss Clarke dies at 76—Retired Engineer at GE," *Schenectady Gazette*, Nov. 19, 1959. "Biographical data— Edith Clarke," The above listed items are all from the Edith Clarke File. Also see an obituary in *Elec. Eng.*, vol. 79, p. 108, 1960, and *Who's Who in Engineering*, p. 439, 1954.

10

1946 ## Women Can Be Engineers

Alice C. Goff

By an irony of fate, war, always bitterly denounced by women, has advanced them in the engineering profession.

Prior to World War I, a few women pioneered in applied science. But it was during that Great War, when men were not available, that the services of women engineers were in great demand. They were drawn into tool design and into chemical research. They designed buildings and automobiles. Great numbers of positions were open to women to carry on professional work formerly done by men.

After the first World War had brought these women into prominence and they had demonstrated their ability, other women became interested in engineering and applied science. As a result, increasing numbers have matriculated in engineering schools and institutes of technology. Many women have received their bachelor's degrees in their chosen subjects. Some have gone on to earn their master's degrees. A few possess doctorates.

This awakened interest in engineering has not been limited to women in the United States. In other parts of the world women have studied the science by which the mechanical properties of matter are made useful to man in structures and machines. In fact a more decided trend in this direction has developed in some other countries than

Goff, Alice. (1946). *Women can be engineers*, Youngstown, Ohio. Excerpts.

the United States. So many British women engineers did important work during World War I, that the Women's Engineering Society was incorporated in London in 1920, and has become a strong organization. Out of the Russian revolution came a changed attitude toward women's work. Russia had been a nation of approximately 90 percent illiterates. In the Soviet Union, women flocked into the scientific and engineering professions. Likewise in Turkey, the revolution wrought great changes in women's work. Turkey had long held the custom of seclusion for women. After the revolution, women left the harems; discarded their veils, which had hidden their faces; and entered the professions. In Canada and Czechoslovakia a few women became engineers.

During the depression following World War I, the path of women engineers was not easy, in Great Britain and the United States. These pioneers fought an uphill battle, for the public has been skeptical of the success of women in engineering. Enough prejudice has existed against women in professions in the United States to force some women engineers to become better trained than men, to obtain similar positions. Likewise in Britain, many women engineers have struggled to secure equal footing with men in their profession.

In marked contrast to the attitude of the public toward women engineers in the United States and Great Britain, was the position of women engineers and scientists in Russia and in Turkey. In those countries no prejudice existed against women in the professions. Women were limited only by their training and their ability. As a rule, the Russian and Turkish women were not so well trained or so advanced in applied science as women of the United States and England. This was not surprising to one familiar with their backgrounds.

In the period before World War II, women engineers and scientists made notable contributions to their professions in various countries. Their achievements became of great value to the war effort of the Allies, as well as to civilian life. Then World War II drew more women into the engineering profession than ever before.

The writer, herself, is a structural engineer, who designs reinforced concrete buildings. She and other women engineers await with interest to see how World War II will affect the status of women in their profession in the United States. The war is certain to result in notable gains for women engineers and scientists, because war speeds up forces that have long been in operation.

The examples selected are a few of the outstanding women engineers and scientists. This book does not profess to cover the entire field.

[. . .]

Olive W. Dennis [1885–1957. Ed.]

Construction work always fascinated Olive Dennis as a child, so her later ambition to become a bridge engineer was the logical outgrowth of her young ideas. When she was ten years old, she built her little brother a trolley car, that he considered a triumph. It was an open car with a pivoting trolley pole, reversible seats, a movable footboard, and a fender that could be changed from one end of the car to the other.

Olive's parents brought her up with extreme care. She was taught to cook and sew, but says that she handled a needle like a marlinspike. A thimble was never of use to her. Although naturally left-handed, she acquired the ability to write with her right hand when she went to school. As a result, she was ambidextrous in her sewing. At a corner of her material she was apt to change the needle to whichever hand seemed convenient. In the shift, the

Olive Dennis.
Courtesy of the B&O Railroad Museum, used with permission.

thimble was usually stranded on the opposite hand from the one using the needle. The art of sewing was mastered only by determined perseverance, but Olive soon became proficient in driving nails, in using saws and chisels, and in turning screws into place with screw drivers.

Sometimes she was late for school, because she lost all sense of time in the fascinating occupation of watching a building under construction. Her mother insisted that she have dolls for amusement; and that she should not hobnob with carpenters and bricklayers. Olive was fairly bursting with questions she wanted to ask about construction work. But she met her mother's requirements to the letter. She watched workmen in silence from a discreet distance. Later she imitated their various skills. Instead of making clothes for her dolls, she built doll-houses and made furniture.

At the age of eleven, Olive built a playhouse. An old woodshed had been torn down and replaced by a new one in the Dennis yard. Olive secured permission from her father to use the old lumber. The building which she erected from it was a finished product with three windows and a door, a sloping roof, a front porch, and three steps. It even boasted wooden shutters and shingles. The playhouse was high enough for the young builder to stand erect inside, under the high side of the roof. At that time she knew nothing of dead and live loads, or the stresses that they produce in various members of a building. However the framing was in careful imitation of the best styles of carpentry that Olive had observed.

As she grew older, she was fascinated by the engineering profession in general, and by structural engineering in particular. If given a choice, she would have chosen to be builder of the Delaware River Bridge in preference to President of the United States.

Since she had been educated in the Baltimore public schools, she decided to go on to Goucher College. There she majored in mathematics and science. From Goucher College she received her bachelor's degree, and a fellowship in mathematics at Columbia University. At the latter institution of learning she majored in mathematics and astronomy, and was rewarded with her master's degree. Then she was appointed as teacher of mathematics in the McKinley Manual Training School, a technical high school in Washington, D.C. She had achieved scholastic distinction, for she had become a member of Phi Beta Kappa.

During the time she was teaching, the idea of studying civil engineering persisted in the back of her mind. So she attended two summer sessions of the engineering school at the University of Wisconsin while on vacation from teaching high school. One of her courses was surveying, which she selected to help her in teaching trigonometry. During the winters she continued with extension courses in civil engineering. Then she resigned from her teaching position in favor of a full year at the engineering school of Cornell University. She completed the requirements for the degree of Civil Engineer at Cornell in 1920. She had specialized in structural engineering.

In the fall of 1920, Miss Dennis secured her first engineering position in the bridge department of the Baltimore and Ohio Railroad at Baltimore, Maryland. As a draftsman, she had an opportunity to apply the theories she had learned at college. In this practical building work, she found that railroad bridges were carefully designed, not only to carry their own dead loads and the live loads imposed on them by trains pounding over them, but for various emergency loadings also. Allowance was made for horizontal pressure due to wind, for vibrational cadence caused by hammering of locomotive drivers, for centrifugal force on the outer rail in rounding a curve, and for expansion and contraction of metal parts due to variations in temperature. Provision was made for impact, braking and tractive forces.

It just happened that at the time Miss Dennis assumed her duties as engineer, the late Mr. Daniel Willard, then president of the Baltimore and Ohio Railroad, was seeking a woman with technical training to make a survey of the whole system and to suggest ways of improving the service. Inasmuch as approximately half of the passengers on the trains were women, he felt someone on his staff should represent the woman's point of view. He believed that technical training was essential, for the woman observer should be able to understand practical requirements of car construction and of railroad operation and maintenance. Mr. Willard had found from experience that persons equipped with technical training were more apt to look for causes underlying certain conditions and reason them through to logical conclusions than others with no technical knowledge.

Miss Dennis had ideal qualifications for such a position, but she was very reluctant to sever her connections with the engineering department. She felt that she must prove to herself, as well as to her skeptical friends, that she could make practical application of her theories about concrete and steel construction. After fourteen months of work, she had demonstrated that she could do it successfully. Some of her ambition to be a practical builder had been satisfied.

In the fall of 1921, she accepted promotion to her present position of engineer of service for the Baltimore and Ohio Railroad. She undertook the assignment from the president's office, on condition that she might return to the engineering department if she wished. With much fear and trepidation, she reported to start her new duties. Her uneasiness was due to the vagueness and generality of her assignment. She had been instructed to ride as a regular passenger on trains of the Baltimore and Ohio Railroad, and other lines; to observe the service carefully and to suggest ways of improving it. In the absence of Mr. Willard, she turned to Vice-President James S. Murray for help in planning her first week's itinerary. He proved so helpful and kind that she adopted him as official adviser in important matters. Soon she became accustomed to her new responsibilities and to her wonderful opportunities for originality and invention. She was forced to admit that she really enjoyed her work.

Because of her decision to undertake this constructive work as engineer of service, former luxuries of travel have become commonplace. In the transition during the past two decades from hot, malodorous, dirty, day coaches with uncomfortably high seats and poor lighting, to modern air-conditioned day coaches and lounge cars, with comfortable seats and improved lighting, a large part of the credit for the transformation is due Miss Dennis. She was responsible for the disappearance of the old-fashioned, glaring red and green plush seat-covers; and for their replacement by soft mauves, sea greens, mulberry tans, and subdued blues—all so restful to the eyes. She suggested the first reclining-seats with comfortable foot-rests and fresh linen seat-backs, for Baltimore and Ohio coaches. Many conveniences which were established following her recommendations were promptly adopted by other railroads.

From the start, she spent half of her time riding on various trains to observe the service. The other half was spent in the office at her home, where she worked out problems she had encountered on the road, and typed confidential reports to Mr. Willard. The suggestions contained in her reports were passed on by Mr. Willard to various officers of the road for consideration. Many of her suggestions were adopted in the course of time. Sometimes it took years of time and many repetitions of a suggestion in varying forms, for an idea to bear fruit. That procedure was natural, because changes in railroad equipment and practice were too costly to be made lightly. This conservative measure was the balance wheel that safeguarded against the adoption of too radical ideas. Miss Dennis found that most of the Baltimore and Ohio executives were receptive to new ideas and were willing to give them fair consideration. They were anxious to pioneer in new developments, which would make the Baltimore and Ohio Railroad the best in point of service in the whole country. Some of her reports received special endorsement by Mr. Willard, to facilitate adoption of her suggestions. It is true, she was handicapped by the attitude of a few officials, who hesitated to reconsider an idea that had failed once, even though basic conditions were entirely different.

During the first two years, Miss Dennis rode on tickets to obtain the regular passenger's point of view. Her ignorance of general railroad practice was an asset, because her reactions were the same as those of the occasional traveler. She made the same mistakes in interpreting ambiguous time-tables that a stranger would. Therefore, she was in position to detect weaknesses and suggest improvements. In buying tickets, in checking baggage, and in locating trains, she received the same impressions from certain features of the service as others would. As a result, she suggested methods that were simpler for the passenger.

In undertaking the work for Mr. Willard, she had the understanding that she was not to act as a "spotter" or spy on individuals. Her suggestions of improvements in service were to be made without selection of individual employees for criticism or commendation. In more than twenty years of service, Miss Dennis has followed this practice as consistently as possible.

As she acquired experience in traveling, she lost much of the freshness of her original viewpoint, with its early advantages. For in the first year alone, the mileage traveled by her reached the amazing total of 44,000 miles. In subsequent years she has continued to travel sufficiently to keep well posted on all the trains rolling back and forth over the 7,000 miles covered by the Baltimore and Ohio Railroad systems, as well as those of immediate competitors, and the "crack" trains throughout the country.

When she learned more about limitations imposed by agreements with labor unions and with other roads, she gained much through acquaintance with various departments of the Baltimore and Ohio Railroad. By consulting directly with department heads before she submitted written reports, she improved her methods of presentation and secured closer cooperation.

In her investigations of service on trains, Miss Dennis has concentrated on day-coach service. Only by riding considerable distances in the lower-priced accommodations on the less-featured trains, both by day and night, could she really discover the little details of service which might be improved for the greater comfort of the passengers. There was little necessity for riding in luxurious Pullman rooms of the finest trains, because that phase of railroad service required little observing. The day coaches tended to express the individuality of each particular railroad more than the Pullmans, which were by agreement of identical standards on all lines. To forge ahead in the days of keen competition in transportation, a railroad was forced to attract its patrons by the quality of service offered them. The forward-looking president of the Baltimore and Ohio Railroad system realized that the great bulk of potential passengers, whom he wished to attract to his railroad by superior service, was composed of persons of moderate means. They were forced to travel economically, and naturally chose the best transportation available, within their budgets. Mr. Willard actually foresaw the present era of "deluxe" coach trains, back in 1921. Therefore, he instructed Miss Dennis to concentrate on investigation of coach service.

Her first rides in day coaches in 1921 resulted in reports regarding three major discomforts, which had to be rectified. Many coach seats were too high for the average woman passenger, so her feet dangled without support. Most coaches were very dirty, because standards and methods of cleaning had declined when the railroads were under government control during the rush of World War I. Finally, the coaches were actually malodorous from inadequate ventilation, especially in winter.

Partial solutions of these three important problems were adopted at that time. The standard height of the walk-over seat was lowered. Old methods of blowing dirt into seat cushions were abandoned. New suction methods were used to pull the dirt out of the cushions. Auxiliary car-cleaning systems were organized, without waiting for the regular periods of reconditioning in the shops. The greasy, grimy plush at the tops of the seat backs was scoured at certain intervals with soap and water and special cleansing preparations. The number of ventilators in the deck sash of the cars was doubled. In order to obtain maximum efficiency from the limited equipment, crews were more carefully supervised in the control of ventilators and in the regulation of steam heat.

Miss Dennis was so impressed by the importance of adequate ventilation that she designed and received a patent for an individually-operated ventilator. This small contrivance was installed in the dash under each window pane on the Baltimore and Ohio trains. The shutter device formed an adjustable aperture for the inlet of fresh air. Any passenger was thus enabled to introduce a comfortable breath of fresh air in even the coldest weather without drafts on the occupants of the seat behind. This arrangement was particularly helpful when summer rains compelled the closing of the regular car windows.

During the summer of 1928, Miss Dennis reported on the terrific heat in the interior of the cars for the National Limited, after they had stood exposed to the hot sun in the yards at St. Louis, before their noon departure. She had observed that refrigerator cars were chilled in advance, in order to be effective in keeping perishable foods at safe temperatures. After some thought on the subject, she concluded that passenger cars might have ice-chilled air blown into them before the passengers entered. The resulting discussion led to the development by the Baltimore and Ohio Railroad of pre-cooling machines for use at all important terminals in 1930.

Miss Dennis had long been advocating improvements in ventilation; she had invented ventilators, and recommended pre-chilled passenger-cars. The inevitable result was that her railroad pioneered in the development of air-conditioning systems, with their automatic control of heat and humidity, for passenger trains. The first practical, mechanical, air-conditioning system for railroad cars was worked out through the ingenuity of the mechanical officers of the Baltimore and Ohio Railroad. Not content with the first results, these men continued to experiment along this line for several years, and were constantly improving the system. The Baltimore and Ohio ran the first air-conditioned car in regular service, the diner Martha Washington in 1930; the first completely air-conditioned train, the Columbian in 1931; and the first fully air-conditioned long-distance sleeping-car train, the National Limited in 1932. The adoption of air-conditioning for trains spread promptly to other railroads.

Meanwhile in 1925, severe competition from bus transportation forced further improvement in railway coach seats. Miss Dennis made an analytic comparison of various features of the motor coach and the railroad day coach. Even with the inclusion of the many improvements that the Baltimore and Ohio Railroad had made in its day coaches, she found that some features were still inferior to those in bus construction. She reported: "Bus chairs are soft and comfortable. The seats seem an inch or two lower than railway coach seats, and are inclined slightly so that the back is a trifle lower than the front edge. The passenger leans more comfortably against the back support." Seat manufacturers had previously insisted that reversing necessary for railroad coach seats required movable backs. These could not be built to support the lower part of the spine as comfortably as an overstuffed chair. Mr. Willard was determined that this obstacle should be surmounted.

Finally by working in conjunction with the Baltimore and Ohio mechanical department, a seat manufacturer succeeded in producing the first double revolving seats. These "individual" or "bucket-type" seats were introduced in Baltimore and Ohio coaches in January 1926. They were used for several years on that road before they were adopted by other railroads.

In the summer of 1930, the Baltimore and Ohio Railroad pioneered in the use of special-feature coach trains for night service between New York and Washington. Many conveniences advocated by Miss Dennis had been installed, when the coaches were remodeled in the company shops at Baltimore. This marked the first time that all these extra comforts of service had been embodied in regular coaches on any railroad. Incidentally, this was four years before the introduction of any of the streamliners, which now carry most of these features. For the remodeled coaches, the Baltimore and Ohio had developed the first rotating seats with reclining backs, adjustable in the manner of a Morris chair. They were neat-looking and comfortable, with their linen covers on seat backs, and their foot-rests. Such seats are now considered standard equipment for new coaches on most American railroads. The ceiling lights in the body of each car had been equipped with rheostats, so that the lights could be dimmed at night to allow coach passengers to sleep undisturbed by glare. Sidewall bracket lamps had been installed at alternate seats, to provide individual lighting for those passengers who wanted to read before falling asleep, without disturbing their neighbors. Tickets were checked through, so that the passengers need not be disturbed during the night. Novel features introduced in these cars included inexpensive lunch counters, open for service all night. The items sold included sandwiches made to order, tea, coffee, and soft drinks. Each item was sold for the nominal sum of five or ten cents. The lunch counter attendant served in triple capacity, for he also handled baggage for passengers, and kept lavatories in order. The coaches were equipped with spacious dressing rooms for women. These were particularly helpful to mothers with small children. The dressing rooms were improved by the addition of an abundance of mirrors, with one full length door-mirror in each. These rooms were provided with hot and cold running water, paper towels, liquid soap, and paper drinking cups. Each dressing room had a separate toilet annex.

Even after her recommendations had been adopted by her road, and had spread to other lines, Miss Dennis continued her inspections of service. By riding on coaches night after night for thousands of miles, she was able to check up on handling of lights and ticket collections. In this follow-up work, she has become somewhat of an authority on sleeping sitting up. Strenuous as her job has been, it has proved worth while. Nowadays, soldiers and civilians who make long treks by rail are indebted to Miss Dennis for many comforts of travel.

In the spring of 1926, it was learned that the contract of the Baltimore and Ohio Railroad for the use of the Pennsylvania Terminal in New York was not to be extended. Mr. Willard asked Miss Dennis to work up some suggestions for a motor coach connection for transfer of passengers from Jersey City into Manhattan. Her report contained recommendations for location of terminals, coach routes, types of motor coaches, handling of passengers and their baggage. Her report served as a basis for the work of the committee of officers, who put into operation the present plan of transfer, direct from railroad cars to motor coaches.

During the first arrangements for the sale of combination rail and airplane tickets in 1928, Miss Dennis was sent to try out the service by air from Baltimore to the Twin Cities, and later to other points. In each case she was air sick, so was excused from further experimentation up in the air.

A project, in which she was given a free hand in remodeling and redecorating of equipment, involved the conversion in the company shops of two day coaches into lounge cars. In May 1929, these lounge cars were needed for immediate use on the Columbian. In fact, a time limit of thirty days had been set, so the furnishings available were restricted. But Miss Dennis seized the opportunity to demonstrate her theory that the interior of a lounge car should resemble a room in a club or a home. It should not be so obviously just a railway car, with straight rows of chairs of identical shape and size in stiff arrangement along the side walls. These were the first cars of this description on Eastern lines to be furnished with settees and chairs of different types and sizes; or with chairs upholstered in different fabrics of contrasting bright colors. This innovation was so enthusiastically received that the *Railway Age* carried a description and illustration of these cars. After the lounge cars of the Columbian were developed with charming roomlike interiors in 1929, the lounge cars of the Columbine and Portland Rose on the Chicago and Northwestern and the Union Pacific Railroads were similarly treated in 1930. In 1934 the streamline trains appeared, with contrasting colors used in their interiors.

The subjects investigated by Miss Dennis and outlined in her reports were diverse, and representative of different lines of activity. Probably none was further removed from her training as an engineer than dining car service. She did not allow her previous education to deter her from undertaking a thorough study of the subject of dietetics and nutrition. After sufficient research to secure authorities for backing on her ideas, she worked out menus for lighter table-d'hote meals, with proper combinations for a balanced diet. Emphasis was placed on salads to please the woman passenger who was interested in her figure. Surprisingly, when the lighter meals were offered on the menu cards, they proved popular with men as well as women passengers. The trend in restaurants throughout the country was toward substitution of lighter dishes in the heavy meals that had prevailed; but Baltimore and Ohio dining cars were among the first railroad cars to follow the trend.

Not only menus, meal service, and table setting concerned Miss Dennis, but she went so far as to design and patent the attractive blue colonial china provided in the dining cars. Three years were required to complete this task. The special pictorial china, suitable for use in the colonial dining cars, embodied pictures of historic Baltimore and Ohio motive power and scenes along the road. A number of dinner plates were prepared to sell as souvenirs of the Baltimore and Ohio Centenary Celebration in 1927. During the three weeks of the Fair of the Iron Horse more than 2000 orders were taken for pieces of this china to be shipped to all parts of the world. The public bought not just plates, but whole sets of dishes. Many more orders were taken during the Century of Progress at Chicago in 1933 and 1934, and continue to come in to the dining-car department.

In the broad field of her work, Miss Dennis has had considerable experience in preparing material for descriptive booklets. One of these was for the new blue china. Others concerned the Baltimore and Ohio right-of-way. Some had involved much historical research. Shortly after the Centenary Celebration in 1927, she undertook to locate the sites of blockhouses erected

B&O Railroad china.
Courtesy of the B&O Railroad Museum, used with permission.

along the Baltimore and Ohio lines during the Civil War for the use of the federal troops guarding bridges and other strategic points. This project entailed much research in libraries, during which Miss Dennis examined thousands of photographs of the Civil War made by Gardner, Poe, Brady, and others. After she had received a list of the sites of these blockhouses from the Old Records Bureau of the War Department, she spent several weeks at bridge sites along the Baltimore and Ohio line in the hills of West Virginia. There she searched for physical indications of earthworks thrown up around those fortifications. In some instances the outlines of the earth mounds could still be seen. With the aid of a number of older residents of the various localities, Miss Dennis was able to identify the exact sites of all 23 of the blockhouses on the Cumberland division. A committee was appointed by the president to inspect the sites she had located; then to prepare plans for the reproduction of one of these blockhouses on the most suitable site. Markers were to be erected at other historic points along the disputed Valley of the Potomac. The plan has not been completed, but the information was secured just in time. The last of the old timers interviewed has since passed on, so first-hand identification would no longer be possible. The information is secure in the reports, illustrated with maps and photographs.

Along with her other duties, Miss Dennis has done personnel work among women employees of the Baltimore and Ohio system. One of her assignments early in 1923 was a complete survey of working conditions of these women. Particular emphasis was placed on offices in which women were employed at outlying points. Women had been taken on in large numbers during the war years, even in offices not originally intended for use by women. The management desired that these employees should have all necessary facilities for privacy in wash rooms, and clean, comfortable working quarters, without unnecessary luxuries. Miss Dennis visited all points of the system where women were employed, and reported on the working conditions of the 2,000 women employees.

During the depression years, she helped with organization of the employees in the Cooperative Traffic Program Committees. She was elated to find her own enthusiasm for the railroad reflected in the spirit of many other employees throughout the company. They felt genuine pride in the history of the Baltimore and Ohio Railroad, and its accomplishments. This feeling was disclosed by the service rendered the public.

In the fall of 1925, a music club was organized in Baltimore to further friendly relations among women of the Baltimore and Ohio system. The club has been composed of women employees, or women members of families of employees of the company. The membership has been built up to 65 under the leadership of Miss Dennis, the club president for the first 17 years of its existence. Two public concerts have been presented annually in Baltimore; one at the Christmas season and the other in the spring. Special concerts have been given on numerous occasions, in Baltimore and at other points on the system. In 1927 the chorus introduced the annual custom of singing Christmas carols in the passenger stations and on the various floors of the general office buildings of the Baltimore and Ohio in Baltimore. Miss Virginia Blackhead, the founding director, trained the singers for the first 16 years of the club's existence. She brought the club from a feeble beginning to its present status as a group of talented singers, who thrill great audiences by their inspiring performances.

Miss Dennis has found working for a railroad a fascinating game, because she has pioneered in a field not ordinarily associated with the work of an engineer. The character of her suggestions for improvements in service has been unrestricted, for she has been unhampered by precedents and red tape. Her work has touched many phases of railroading. Among her many duties, she has made analyses of traffic conditions and business trends.

Since 1927, she has been identified with the American Railway Engineering Association. She has taken an active part in the work of the committee on economics of railway location and operation.

She has many diversions. An expert at many types of puzzles, she prefers cryptograms. She is enthusiastic about Quota Club International, a classified service club for women; and she has served as president of Baltimore Quota. In her spare moments, her garden comes in for much attention.

For her pioneering work in the improvement of railroad service, not only on the Baltimore and Ohio, but reflected in other roads, Miss Dennis has received many honors. At the Women's Centennial Congress in New York in 1940, she was named by Carrie Chapman Catt as one of the 100 outstanding career women in the United States.

[. . .]

Margaret Ingels [1892–1971. Ed.]

Margaret Ingels' interest in air conditioning started when she was a tot on her mother's knee. When she observed moisture collecting on a cold glass,

Margaret Ingels.

her curiosity about the common scientific principle of condensation became insatiable.

During her school days in her home town of Paris, Kentucky, Margaret resolved to learn more about scientific and engineering subjects that fascinated her. This determination carried her on through the engineering school at the University of Kentucky. In 1916 she was granted the degree of Mechanical Engineer. Incidentally, she was the first woman in the United States to be awarded that degree. *[Actually, as discussed in Chapter 8, Bertha Lamme received a degree in mechanical engineering in 1893. Ingels was the first woman to receive an engineering degree from the University of Kentucky. Ed.]* After three years of practical experience and a thesis accepted by the graduate school, she obtained a master's degree in engineering from her alma mater.

With an education so unusual for a woman, she was inspired to repay her privilege by service to mankind. She was afforded an opportunity to begin this outstanding work, when she became identified with the Carrier Corporation in New York. From 1917 to 1921, she had four years of valuable experience with the pioneer air conditioning company, whose products have had a profound effect upon commerce and industry. She was in on the ground floor, for air conditioning was a new art. Its first development in the industrial field had come in 1903, only 14 years previous to her entrance.

By devoting herself to research work for the next eight years, Miss Ingels made herself an authority on problems connected with air conditioning. The research laboratories of the American Society of Heating and Ventilating Engineers at the United States Bureau of Mines in Pittsburgh called her. From 1921 through 1926, her studies included atmospheric dusts, physiological reactions of human beings in various air conditions, and infiltration of air through building walls and around building openings. Several of her papers on these subjects were published in the transactions of the American Society of Heating and Ventilating Engineers, during these five years. Then she undertook field research work in Syracuse, New York, for the New York Commission on Ventilation. Her study was to correlate health and attendance of school children to various types of ventilating systems.

Inevitably she was drawn back to the pioneer air conditioning company, which had afforded her valuable engineering experience. Since 1929, Miss Ingels has been employed by the Carrier Corporation of Syracuse. Her training and experience had provided her with the ability to pass on all engineering information originating in this company. In her present capacity of Engineering Editor, she has written many articles on air conditioning for magazines and technical publications such as: *Industrial and Engineering Chemistry, Heating and Ventilating Magazine, Refrigerating Engineering,*

Manufacturer's Record, Commercial American, Showman's Trade, Real Estate, and *Good Housekeeping.* Her lectures on air conditioning have been attended by thousands throughout the United States. Her clear, nontechnical language has been very effective in the widespread demand for air-conditioning equipment, for commercial and industrial purposes.

She has defined air conditioning as the scientific control of the variables of humidity, temperature, and motion, of cleaned air within enclosures, to produce indoors the atmospheric condition most desired. The general method of accomplishment is by means of special treatment of air circulated within and supplied to the enclosures. Therefore, air conditioning is accomplished through a combination of air cleaning, heating or cooling, humidifying or dehumidifying, and effective control of air motion within the enclosures.

In reviewing the rapid growth of air conditioning in the industrial field, Miss Ingels has stated that the textile mill served as the proving ground for much of the development of air conditioning. Even the term "air conditioning" was taken from the textile industry which used the expression "yarn conditioning" to signify the moisture condition of the fibre. Textile mills required high relative humidity; that is, a large amount of moisture contained in the air at a certain temperature as compared with the amount the air would contain if saturated with moisture. The desired relative humidity of 75 percent was essential in order to maintain a suitable moisture content in the hygroscopic fibre. This moisture content was of utmost importance, because it greatly increased the strength of the yarn and improved its quality. In turn, production was markedly increased on account of reduction of breakage in spinning.

The effects of installations of air conditioning in cotton mills were far reaching. Controlled indoor weather was an important factor in the moving of the textile industrial centers from New England to locations close to the cotton fields. Air conditioning has made possible the change of an agricultural South to an industrial South.

Although textiles absorbed the attention of air conditioning in its infancy, applications rapidly spread to other industries. The great tobacco industry of the South used a large quantity of conditioned air. Many companies utilized it throughout the entire process of manufacture, from the storage of hogshead to the storage of the water-proof wrapped package. Tobacco broke and cracked when exposed to the dry air. Therefore, air conditioning was important to the industry, from the standpoint of maintaining high, uniform quality of product, and of meeting manufacturing schedules.

Miss Ingels, with her southern accent and charm, and with her loyalty to her native Kentucky, must derive satisfaction from the fact that air conditioning has played such an important role in the industrial South. Great, diversified industries of the South are dependent upon manufactured indoor weather. Rayon cannot be made without air conditioning; and many of the plants manufacturing rayon are below the Mason and Dixon line. *[. . .]*

Miss Ingels pointed out the fact that in very dry climates at high altitudes, where the relative humidity is always low, evaporative cooling may be obtained in homes without excessive addition of water. This principle was not discovered recently, for it was employed by the Mongol conquerors of India centuries ago. Water fixtures attached to the ruins of their deserted palaces testify to this early practice. In fact, this same principle of evaporative cooling is used in India today. Wetted grass mats are placed over openings on the windward side of the palaces. The air blowing through these mats cools the interiors considerably below the outside temperature. During the extremely hot dry season, the difference in temperature often amounts to 20 degrees.

Miss Ingels has outlined the air-conditioning equipment required for the control of temperature, moisture, motion, and cleanliness of air. For heat addition, she has listed the interchanger and the heating coil. Hot gases from combustion, hot water, or steam, may be used as the heating medium. When the products of combustion are utilized as the heating medium, the gases are passed through an interchanger. The air to be heated passes over the gas compartments. However, when hot water or steam is the heating medium, it is circulated through a heating coil. The air to be heated passes over the coil.

For the removal of heat and moisture, she has listed the cooling coil and cold spray as equipment extensively used. Air may be cooled and dehumidified by passing it over cooling coils, through which cold water or a refrigerant is circulated. The same effect may be secured by passing the air through a cold spray. Another method of moisture removal is to pass the air over a hygroscopic chemical which will absorb the moisture.

Miss Ingels has stated that the equipment for moisture addition may be a pan humidifier, target spray, or spray nozzles. The simplest method is to place a heating coil below the surface of the water in the pan humidifier, to secure ready vaporization. For small installations, a jet of water may be shot against a target to cause the fine spray to be evaporated. But for exact means of moisture control in industrial installations, spray nozzles are most accurate. They are arranged in banks, and are placed perpendicular to the direction of air flow. Air, passing through the sheets of water formed by the nozzles, leaves the spray chamber saturated at the temperature of the leaving water.

For air motion, a fan is a required part of the air-conditioning system. Its design depends upon the quantity of air to be circulated, the over-all resistance to air flow, the maximum noise level, the minimum power consumption, and proper balance to reduce vibrations.

For cleanliness of air, filters and washing are used to remove dusts. There are many combinations of the component parts of air-conditioning apparatus.

In systems of the large central station type, the apparatus is centrally located and conditioned air is distributed through a system of duct work in or near ceilings of spaces to be conditioned. Air at the floor is drawn into return ducts, for buildings such as theatres. In systems of the small unitary type, relatively little duct work is required for air distribution. For the latter, equipment is assembled at the factory in standard size units.

Soon after the entrance of the United States into World War II, house building tapered down to practically nothing. Construction of theatres, department stores, hotels, office buildings, and restaurants, gave way to building of plants and factories required for exigencies of war. From the start of the war, the Carrier Corporation had done war work almost exclusively. Miss Ingels found her work more vital than ever before.

In fact, air conditioning has played an important role in many industries essential to successful warfare. Precision instruments, which must be accurate to 1/10,000th of an inch, can be made only in rooms where humidity, temperature, and dust are controlled. For ammunition plants, certain types of ammunition must be made in air-conditioned factories to guard powder and fuses. Nylon fabrics require complete control of atmospheric conditions in their manufacture. Aircraft plants were air conditioned in order to make finer engines. Controlled weather made new products possible in the field of synthetic rubber. Most important of all, air conditioning allowed bombsights of amazing accuracy to be manufactured with greatest precision. Constantly uniform temperature prevents expansion and contraction of optical glass, and so permits grinding and polishing to the nearly incredible tolerance of 6/1,000,000th of an inch. Removal of dust and grit from the air protects delicate surfaces from damage. Humidity control prevents damage to costly materials from moisture, and rusting of metal parts and tools.

Miss Ingels thoroughly appreciates the progress in machines resulting from intensified war work. She realizes that these advances in war time may later be applied to production in peace that follows. She predicts a million houses will be air conditioned; and that factories and offices generally will have controlled indoor weather. She even talks of air-conditioned cities as a possibility of the future, for such predictions have been considered by Dr. Willis H. Carrier, founder of the science of air conditioning.

The importance of Miss Ingels' work in the air conditioning field has been recognized in various ways. Her work is cited in the current edition of the *Encyclopedia Britannica*, which mentions her findings on the value of the Kata Thermometer in effective temperature studies. No other woman engineer in the world has been accorded this distinction. She was honored by the Women's Centennial Congress in 1940, when she was selected as one of the pioneers in the evolution of women's careers in the past century.

Miss Ingels was responsible for an ingenious device, called an "igloo," for air conditioning feet at the New York World's Fair. She had journeyed over the concrete paving in the heat of August, until she became footsore and weary. Her feet were like hot coals. This experience inspired her to sketch a throne to hold three people simultaneously. The result was that such a device became a part of the Carrier display at the fair. In front of each seat was located a square grill through which air at 40° F flowed at 300 feet a minute and was circulated by means of suction fans. One went on the stand, removed his shoes, and cooled his feet. The apparatus proved immensely popular at the fair.

Miss Ingels is engaged in many activities. She is a member of the American Society of Heating and Ventilating Engineers, the American Association of University Women, and the Technology Club of Syracuse. She is an expert bridge player, and an enthusiastic golfer. She is as proud of her skill in housekeeping and knitting as of her knowledge of logarithms and thermal units.

Her favorite spot is Kentucky. In connection with hundreds of lectures she has delivered, she has checked the relative humidity of air with a gadget know as a sling psychrometer. In the only instance that she found the atmosphere perfect without air conditioning, she happened to be in her native Kentucky.

Elsie Eaves [1898–1983. Ed.]

Elsie Eaves received her early formal education in the grade schools of Idaho Springs, Colorado; and later in a high school there. Even then she excelled in mathematics, so she computed mining surveys for Walter Funk, home town mining engineer.

Her fondness for mathematics inspired her to undertake a course in civil engineering at the University of Colorado. However, she began her engineering experience before she received her university degree. She was a draftsman for the United States Bureau of Public Roads at Denver, and then for the Denver and Rio Grande Railroad Company. In her senior year she was secretary to Dean Herbert S. Evans of the college of engineering. In 1920 she was awarded the degree of Bachelor of Science in Civil Engineering by

Elsie Eaves.
SWE Archives, Walter P. Reuther Library, Wayne State University, used with permission.

the University of Colorado. She remained there as instructor in engineering mathematics the following year.

After a short time with the Colorado state highway department, she was office engineer and assistant to H. S. Crocker, consulting engineer and contractor of Denver. In 1926, Miss Eaves went to New York City to become associated with *Engineering News-Record* and *Construction Methods*. She started as assistant on market surveys. In 1928 she became director of market surveys. Since 1932 she has served not only in that capacity, but also as manager of the business news department.

She is in charge of construction-news gathering through 115 representatives throughout the United States and Canada. The scope of this work may be appreciated from the totals of all construction in the United States for the past few years. In 1938, the total civilian construction approximated $7 billions, including $3 billions of private work and $4 billions of public work. War construction boosted the totals for both public and private work the following year. It accounted for about half of the $8.5 billions during 1940. In 1941, war construction assumed a still larger proportion of the total of nearly $11.5 billions. During 1942, war work rose from 60 to 88 percent to dominate the construction industry. In spite of restrictions imposed on residential building and elimination of all non-essential work, the construction industry experienced its biggest year in history. The volume of construction work reached the staggering total of approximately $12.8 billions.

As manager of the business news department, Miss Eaves is in charge of field collection, and publication of classified construction-news reports in the *Engineering News-Record*. Projects are screened by size. The minimum size of water supply, earthwork, and waterworks projects listed is $15,000; and of other public works, $25,000. The minimum size of industrial buildings listed is $40,000; and of other buildings $150,000. Projects are grouped into various classes of construction, such as: Water Supply, Sewers and Waste Disposal, Bridges and Grade Crossings, Streets and Roads, Earthwork and Waterways, Public Buildings, Commercial Buildings, Industrial Buildings, and Construction in Latin America. Various stages of progress are reported. First, proposed work is listed. Then, dates are given on which bids are asked. Later, the low bidders are listed. On projects below $500,000 value, these are the final reports published, except where award is not made to the low bidder. In such a case, a supplementary contract award report is published. For projects of $500,000 or over, final reports cover contracts awarded. Such projects are starred, so that they may be easily spotted.

Miss Eaves is publication and sales manager of the McGraw-Hill *Construction Daily*. This daily news service covers approximately the same ground as the weekly reports in the *Engineering News-Record*. Since it is issued more frequently, it is of much greater benefit to contractors, engineers, manufacturers, and insurance companies. It affords them prompt information as to what the new construction jobs are, where they are located, who the owners, engineers, and contractors are. It enables them to locate the largest projects and most active territories for business. The listing of projects geographically, as well as according to type of work, and present status, gives the reader a comprehensive picture of the entire construction work of the United States.

The business news department of the *Engineering News-Record* compiles and processes these statistics gathered into the trends, tables, and charts, used in the editorial pages of the magazine. Significant movements in the construction industry are noted. For example, in April of 1942, the statistics already gathered indicated that $12.5 billions would be spent in the continental United States during the year, for construction that does not include materials sent to off-the-continent bases and theater-of-operations construction. The volume of the latter was expected to add $3 billions to the predicted total. In October, 1942, estimates forecast a volume of about $8.2 billions on the home front for United States construction in 1943. Although this volume receded about one-third from the record-breaking year of 1942, it was indicated that substantial gains would be made in out-of-the-country building by the United States. Analysis was made of the distribution of the enormous public-building volume in the defense and war areas. This was made by type of structures: manufacturing plants, mass housing, military buildings, and other public buildings. In April 1945, estimates forecast $20 billions for postwar projects.

The department of market surveys, of which Miss Eaves is director, interprets construction statistics from various sources for manufacturers and distributors interested in market trends and marketing opportunities. These are made available to them through the service bulletin, *Engineering Construction Markets*, edited by Miss Eaves. The market survey work includes the building up of extensive classified records for answering questions from engineers and contractors as to where they can buy products they need. This work became of vital importance in World War II. The armed forces, Army and Navy engineer regiments, and lease-lend shipments, took 85 percent of the 1942 new equipment production. Only 15 percent of new equipment was left for contractors. They were handling an all-time record volume of heavy construction. These men were wearing out their equipment on rush jobs faster than at any previous time in history. The restriction on the new equipment available to contractors, and the punishment taken by the old equipment, together increased the importance of classified records as to where products could be obtained to keep old equipment in service. As shortages of critical materials developed, substitutions presented a problem to contractors; so classified records were essential.

In addition to her work as editor of the service bulletin, Miss Eaves conducts special investigations to determine changes in buying habits and needs of the market. The market survey work also includes preparation of special reports on marketing methods effective in the civil engineering construction market.

All the statistics gathered and interpreted by Miss Eaves and her associates are of tremendous value in times of peace, because construction is America's great peacetime industry. But in World War II, such statistics were of infinitely greater importance and necessity. The whole country was geared up to speed through the war program. Yet the tremendous production desired could not be accomplished without the swiftly executed construction jobs of mammoth size. This construction work which set the stage for production was hastened by the work of *Engineering News-Record* and *Construction Methods*.

Miss Eaves has received many honors. She has the distinction of being the first woman elected to corporate membership in the American Society of Civil Engineers. At the University of Colorado she was initiated into Pi Beta Phi and Hesperia. She belongs to the Colorado Society of Engineers, and the Woman's Engineering Society, London. She is a member and past president of the Altrusa Club of New York.

She takes a keen interest in what other women are doing in engineering. She is the author of the chapter on "Civil Engineering" in *An Outline of Careers for Women*, and also of magazine articles in regard to women engineers. One of her hobbies was "Houseboat Argument," although it was given up for the duration of the war. Naturally, for one so fond of water as to favor a houseboat, her favorite sport is swimming. In spite of the urgency of her work during the war, she has found time to remodel an old house into apartments to rent, in order to finance a future "Houseboat Argument II."

11

1999   *When Computers Were Women*

Jennifer S. Light

J. PRESPER ECKERT AND JOHN W. MAUCHLY, HOUSEHOLD names in the history of computing, developed America's first electronic computer, ENIAC *[Electronic Numerical Integrator and Computer. Ed.]*, to automate ballistics computations during World War II. These two talented engineers dominate the story as it is usually told, but they hardly worked alone. Nearly two hundred young women, both civilian and military, worked on the project as human "computers," performing ballistics computations during the war. Six of them were selected to program a machine that, ironically, would take their name and replace them, a machine whose technical expertise would become vastly more celebrated than their own.[1]

The omission of women from the history of computer science perpetuates misconceptions of women as uninterested or incapable in the field. This article retells the history of ENIAC's "invention" with special focus on the female technicians whom existing computer histories have rendered invisible. In particular, it examines how the job of programmer, perceived in recent years as masculine work, originated as feminized clerical labor. The story presents an apparent paradox. It suggests that women were somehow

Light, Jennifer S. (1999). "When computers were women." *Technology and Culture*, 40(3), 455–483. © The Society for the History of Technology. Reprinted with permission of The Johns Hopkins University Press.

hidden during this stage of computer history while the wartime popular press trumpeted just the opposite—that women were breaking into traditionally male occupations within science, technology, and engineering. A closer look at this literature explicates the paradox by revealing widespread ambivalence about women's work. While celebrating women's presence, wartime writing minimized the complexities of their actual work. While describing the difficulty of their tasks, it classified their occupations as subprofessional. While showcasing them in formerly male occupations, it celebrated their work for its femininity. Despite the complexities—and often pathbreaking aspects—of the work women performed, they rarely received credit for innovation or invention.

The story of ENIAC's female computers supports Ruth Milkman's thesis of an "idiom of sex-typing" during World War II—that the rationale explaining why women performed certain jobs contradicted the actual sexual division of labor.[2] Following her lead, I will compare the actual contributions of these women with their media image. Prewar labor patterns in scientific and clerical occupations significantly influenced the way women with mathematical training were assigned to jobs, what kinds of work they did, and how contemporary media regarded (or failed to regard) this work. This article suggests why previous accounts of computer history did not portray women as significant and argues for a reappraisal of their contributions.[3]

Women in Wartime

Wartime literature characterized World War II as a momentous event in the history of women's employment. In 1943 *Wartime Opportunities for Women* proclaimed, "It's a Woman's World!"[4] Such accounts hailed unprecedented employment opportunities as men were recruited for combat positions. New military and civilian women's organizations such as the Army's Women's Auxiliary Army Corps (WAAC, converted to full military status in 1943 and renamed the Women's Army Corps [WAC]), the Navy's Women Accepted for Volunteer Emergency Service (WAVES), and the American Women's Voluntary Services (AWVS) channeled women into a variety of jobs. The press emphasized the role of machines in war and urged women with mechanical knowledge to "make use of it to the best possible purpose."[5] *Wartime Opportunities for Women* urged: "In this most technical of all wars, science in action is a prime necessity. Engineering is science in action. It takes what the creative mind behind pure science has to offer and builds toward a new engine, product or process."[6] According to the U.S. Department of Labor's Women's Bureau: "The need for women engineers and scientists is growing both in industry and government. . . . Women are being offered scientific and engineering jobs where formerly men were preferred. Now is the time to consider your job in science and engineering. There are no limitations on your opportunities. . . . In looking at the war job opportunities in science and engineering, you will find that the slogan there as elsewhere is 'WOMEN WANTED!'"[7]

A multiplicity of books and pamphlets published by the U.S. War Department and the Department of Labor, with such titles as *Women in War*, *American Women in Uniform*, *Back of the Fighting Front*, and *Wartime Opportunities for Women*, echoed this sentiment. Before World War II, women with college degrees in mathematics generally taught primary or secondary school. Occasionally they worked in clerical services as statistical clerks or human computers. The war changed job demands, and one women's college reported that every mathematics major had her choice of twenty-five jobs in industry or government.[8]

Yet, as Milkman suggests, more women in the labor market did not necessarily mean more equality with men. Sexual divisions of labor persisted during wartime. The geography of women's work settings changed, but the new technical positions did not extend up the job ladder. A widely held belief that female workers would be dismissed once male veterans returned from the war helps to explain the Women's Bureau acknowledgment that "except for Ph.D.'s, women trained in mathematics tend to be employed at the assistant level."[9] The War Department and the Department of Labor actively promoted women's breadth of opportunity yet in some areas explicitly defined which jobs were "open to women." Classified advertisements ran separate listings for "female help wanted" and "male help wanted."

Women's Ambiguous Entry into Computing

Women's role in the development of ENIAC offers an account of the feminization of one occupation, "ballistics computer," and both the creation of and gendering of another, "operator" (what we would now call programmer). Ballistics computation and programming lay at the intersection of scientific and clerical labor. Each required advanced mathematical training, yet each was categorized as clerical work. Such gendering of occupations had precedent. Since the late nineteenth century feminized jobs had developed in a number of sciences where women worked alongside men. Margaret Rossiter identifies several conditions that facilitated the growth of "women's work."[10] These include the rise of big science research projects, low budgets, an available pool of educated women, a lack of men, a woman who could act as an intermediary (such as a male scientist's wife), and a somewhat enlightened employer in a climate generally resistant to female employees entering traditionally male domains. Craving opportunities to use their skills, some women colluded with this sexual division of labor. Many did not aspire to professional employment at higher levels.[11]

Occupational feminization in the sciences fostered long-term invisibility. For example, beginning in the 1940s, laboratories hired women to examine the nuclear and particle tracks on photographic emulsions.[12] Until the 1950s, published copies of photographs that each woman scanned bore her name. Yet eventually the status of these women's work eroded. Later publications were subsumed under the name of the lab leader, inevitably a man, and publicity photographs rarely showcased women's contributions. Physicist Cecil Powell's request for "three more microscopes and three girls" suggests how invisibility and interchangeability went hand in hand.[13] In a number of laboratories, scientists described women not as individuals, but rather as a collective, defined by their lab leader ("Cecil's Beauty Chorus") or by their machines ("scanner girls"). Likewise in the ENIAC project, female operators are referred to as "[John] Holberton's group" or as "ENIAC girls." Technicians generally did not author papers or technical manuals. Nor did they acquire the coveted status symbols of scientists and engineers: publications, lectures, and membership in professional societies. Ultimately these women never got a public opportunity to display their technical knowledge, crucial for personal recognition and career advancement.

Wartime labor shortages stimulated women's entry into new occupations, and computing was no exception.[14] Ballistics computing, a man's job during World War I, was feminized by World War II. A memorandum from the Computing Group Organization and Practices at the National Advisory Committee for Aeronautics (NACA), dated 27 April 1942, explains how the NACA conceived the role of computers: "It is felt that enough greater return is obtained by freeing the engineers from calculating detail to overcome any increased expenses in the computers' salaries. The engineers admit themselves that the girl computers do the work more rapidly and accurately than they would. This is due in large measure to the feeling among the engineers that their college and industrial experience is being wasted and thwarted by mere repetitive calculation."[15]

Patterns of occupational segregation developed in selected industries and job categories newly opened to women.[16] Women hired as computers and clerks generally assisted men. Captain Herman Goldstine, an ENIAC project leader, served as liaison from the U.S. Army's Ballistic Research Laboratory (BRL) to the Moore School of Electrical Engineering at the University of Pennsylvania, which produced ENIAC, and director of computer training for BRL. He recalls that by World War II "there were a few men [computers] but only a few. Any able-bodied man was going to get taken up into the armed forces."[17] With feminization came a loss of technical status, since other men doing more "important" technical and classified work remained in noncombatant positions. Thus, the meaning of "wartime labor shortage" was circumscribed even as it came into being. While college-educated engineers considered the task of computing too tedious for themselves, it was not too tedious for the college-educated women who made up the majority of computers.[18] These were not simply cases of women taking on men's tasks, but rather of the emergence of new job definitions in light of the female workforce.[19] Celebrations of women's wartime contributions thus rarely challenged gender roles. Rather, popular accounts portrayed civilian jobs for women as appropriately feminine, "domestic" work for the nation—despite the fact they were formerly done by men.[20]

The introduction of technology also facilitated women's entry into paid labor. Machines stimulated the reorganization of work processes, often leading to the creation of new occupations and the culling of older ones. In both clerical and factory work, introducing technology changed some jobs so that women performed slightly different tasks rather than substituting directly for men. Women's entry into the workforce was greatest in new occupations where they did not displace men.[21] Once a particular job was feminized this classification gathered momentum, often broadening to include other occupations.[22] By World War II, computing was feminized across a variety of fields, including engineering, architecture, ballistics, and the aircraft industry. The new machines, capable of replacing hundreds of human computers, required human intervention to set up mathematical problems. Without a gendered precedent, the job of computer operator, like the newly created jobs of "stenographer typist" and "scanning girl," became women's work. There is, of course, a fundamental difference between the human computer and the programmer who transfers this skill to an automated process. In the 1940s, the skill of transferring this information—what we now call programming—fit easily with notions about women's work. As an extension of the job of a human computer, this clerical task offered slightly higher status and higher pay than other kinds of clerical labor.[23]

Female Computers and ENIAC Girls

Like much of scientific research and development during World War II, the ENIAC was the offspring of a wartime alliance between a university (the University of Pennsylvania, specifically the Moore School of Electrical

Engineering) and the U.S. armed forces, in this case the Army Proving Ground (APG) in Aberdeen, Maryland. The APG housed the army's Ballistic Research Laboratory (BRL), which produced range tables for gunners. During the war, BRL recruited approximately two hundred women to work as computers, hand-calculating firing tables for rockets and artillery shells. In 1940, when President Franklin D. Roosevelt declared a national emergency, BRL commandeered the Moore School's differential analyzer and began to move some of its work to the university.[24]

One of the first women the army hired to work at the Moore School was twenty-two-year-old Kathleen McNulty. She had graduated in 1942 from Chestnut Hill College, in Philadelphia, with one of the three math degrees awarded in her class. McNulty and her friend Frances Bilas answered an advertisement in a local paper that said Aberdeen was hiring mathematicians:

> I never heard of numerical integration. We had never done anything like that. Numerical integration is where you take, in this particular case . . . [the] path of a bullet from the time it leaves the muzzle of the gun until it reaches the ground. It is a very complex equation; it has about fifteen multiplications and a square root and I don't know what else. You have to find out where the bullet is every tenth of a second from the time it leaves the muzzle of the gun, and you have to take into account all the things that are going to affect the path of the bullet. The very first things that affect the path of the bullet [are] the speed at which it shoots out of the gun [the muzzle velocity], the angle at which it is shot out of the gun, and the size. That's all incorporated in a function which they give you—a [ballistic] coefficient.
>
> As the bullet travels through the air, before it reaches its highest point, it is constantly being pressed down by gravity. It is also being acted upon by air pressure, even by the temperature. As the bullet reached a certain muzzle velocity—usually a declining muzzle velocity, because a typical muzzle velocity would be 2,800 feet per second [fps]—when it got down to the point of 1,110 fps, the speed of sound, then it wobbled terribly. . . . So instead of computing now at a tenth of a second, you might have broken this down to one one hundredth of a second to very carefully calculate this path as it went through there. Then what you had to do, when you finished the whole calculation, you interpolated the values to find out what was the very highest point and where it hit the ground.[25]

The work required a high level of mathematical skill, which included solving nonlinear differential equations in several variables: "Every four lines we had to check our computations by something called Simpson's rule to prove that we were performing the functions correctly. All of it was done using numbers so that you kept constantly finding differences and correcting back."[26] Depending upon their method, the computers could calculate a trajectory in somewhere between twenty minutes and several days, using the differential analyzer, slide rules, and desktop commercial calculators.[27] Despite the complexities of preparing firing tables, in this feminized job category McNulty's appointment was rated at a subprofessional grade. The BRL also categorized women like Lila Todd, a computer supervisor when McNulty started work at the Moore School, as subprofessional.[28]

Herman Goldstine recalls that BRL hired female computers almost exclusively. At first, most women were recent college graduates in the Baltimore and Philadelphia area. Adele Goldstine, his wife and a senior computer, expanded recruiting to include colleges across the Northeast, but the project still needed more personnel.[29] In a short time, recalls Goldstine, "We used up all of the civil service women we could get our hands on."[30] A memo to University of Pennsylvania provost George McClelland from Harold Pender, dean of the Moore School, explained how BRL sought to remedy the situation: "Colonel Simon, Chief of the Ballistic Research Laboratory, has had a specially selected group of WACs assigned to the Laboratory. Although these women have been individually picked they are for the most part ready for training and are not trained persons who can enter fully into the Laboratory's work. . . . By consulting appropriate persons on the campus it appears that this can be carried out without interfering with any of the University's regular work. . . . Under the above circumstances it appears that the University's regular work will not be disturbed and at the same time we will have the opportunity to do a rather important service."[31] Pender's memo embodies a more widespread ambivalence about women's wartime contributions, particularly as members of the military. While "specially selected" for a "rather important" task these women were simultaneously "not trained persons" and could not enter "fully" into the BRL's work.

Colonel Simon assigned two groups of WACs to work as computers. One used desk calculators and the differential analyzer for practical work at the BRL, while the other studied mathematics for ballistics computations at the University of Pennsylvania. These two groups alternated monthly for eight months. The first WAC course started on 9 August 1943. According to reports in the *Daily Pennsylvanian*, the university's student newspaper, these women assimilated smoothly into campus life:

The WACs at present stationed on the University campus are members of two groups alternating in a special course at the Moore School of

Electrical Engineering, and were detached from the unit at Aberdeen Proving Ground, Maryland. At Aberdeen most of them were assigned as computers. The two sections, each of which numbers approximately thirty women, are commanded by second lieutenants and corporals. They are taking courses that are equivalent to the work of a college mathematics major. The results of these studies will later be used in ballistic work at the Ballistic Research Laboratory of the Army Ordnance Department. They are stationed at the Moore School of Electrical Engineering rather than at any other University school because of the large amount of work that the Moore School has done in collaboration with the Ballistic Research Laboratory. They are quartered in the fraternity house [Phi Kappa Sigma], messed in Sergeant Hall, and receive physical training at Bennett Hall. They are required to police their own rooms and be in bed at eleven forty-five P.M., with the exception of weekends. Reveille must be answered at 7:10 A.M.[32]

In this straightforward report, the student reporter neglects to mention the concurrent and widespread tensions surrounding WACs. Only a month earlier, on 1 July 1943, President Roosevelt had signed legislation converting the Women's Auxiliary Army Corps to full military status as the WAC. The conversion was scheduled for implementation by 1 October. According to WAC historian Mattie Treadwell, "The following ninety days of the summer of 1943, initially called The Conversion, were perhaps the busiest in the history of the Corps."[33]

While the article quoted several WACs commenting about their campus lives in a quite positive tone, Adele Goldstine, in an undated letter to a correspondent, reported, "Rumor hath it that the WACs (Sec. I) have been told that they're unloved by everybody including the ES&MWTesses. If it's true, I'm sorry to hear it because I'm afraid it will make our uphill fight steeper."[34] Her letter suggests that the women's presence on campus had become the "interference" and "disturbance" intimated by Simon's memo. Indeed, ambivalence about The Conversion had triggered slander campaigns against WACs from 1943. The cold reception of WAC volunteers was a product not only of news media but also of local gossip: "Resentment was expressed in towns where WACs were quartered, to the effect that they were spoiling the character of the town."[35] The WACs in Philadelphia may have experienced some of the more widespread hostility towards enlisted women.

Separated by skill level into two groups, the WACs at the Moore school had forty hours of classroom instruction per week. According to the syllabus, the course was designed to treat "in succinct form the mathematics which a person should have to work on physical problems such as those likely to be met in the Ballistic Research Laboratory."[36] The mathematics ranged from elementary algebra to simple differential equations. In addition, a unit on the use of calculating machines covered computation- and calculation-machine techniques, handling numerical data, organizing work for machine calculation, and using slide rules.

The instructors included three men (a Dr. Sohon, a Mr. Charp, and a Mr. Fliess) and nine women (Adele Goldstine, Mary Mauchly, Mildred Kramer, Alice Burks, a Mrs. Harris, a Miss Mott, a Miss Greene, a Mrs. Seeley, and a Mrs. Pritkin). Accounts of ENIAC that discuss the WAC course, such as Goldstine's book and the civilian women's own reflections, mention as instructors only three married women: Adele Goldstine, Mary Mauchly, wife of John Mauchly of the Moore School, and Mildred Kramer, wife of Samuel Noah Kramer, a professor of Assyriology at the University of Pennsylvania. Yet archival records show that this is not the full story.[37] Perhaps this oversight is consistent with a different trend Rossiter discusses—that more prominent women in science were often married to notable men, also often scientists. It is unclear whether Goldstine, Mauchly, and Kramer became "visible" because their husbands' visibility accorded them extra attention, because these men somehow facilitated their wives' careers, or because the women themselves campaigned for recognition.

"Thanks for the Memory," a song presumably written by several WACs, offers a playful account of their time at the Moore School:

Of days way back when school
Was just the daily rule
When we just studied theories
For fun and not as tools—thank you so much.

Of lectures running late
Of Math that's mixed with paint
Of dainty slips that ride up hips
And hair-do-ups that ain't—thank you so much.

Many's the time that we fretted
And many's the time that we sweated
Over problems of Simpson and Weddle
But we didn't care—for c'est la guerre!

That Saturday always came
And teach ran for her train
If she didn't lam—like Mary's lamb
Her pets to Moore School came—thank you so much.

Machines that dance and dive
Of numbers that can jive
Of series that do leaps and bounds
Until you lose the five—thank you so much.

Of half-hour luncheon treks
How we waited for our checks!
Of assets, liabilities—
Till all of us were wrecks—thank you too much.

We squared and we cubed and we plotted
And many lines drew and some dotted
We've all developed a complex
Over wine, sex, and f(x)

Of private tête-à-têtes
And talk about our dates
And how we wish that teacher would oblige
By coming late—thank you so much.

And so on through the night.[38]

Even as the WAC courses went on, Moore School engineers were designing a machine to automate the production of the same firing and bombing tables calculated by the human computers: the ENIAC. Engineers wanted answers faster than women could supply them using available technologies. Yet ENIAC couldn't do everything itself. Programming equations into the machine required human labor.[39] The eventual transfer of computing from human to machine led to shifting job definitions. A "computer" was a human being until approximately 1945. After that date the term referred to a machine, and the former human computers became "operators."[40]

Herman Goldstine recounts selecting the operators. At BRL, one group of women used desk calculators and another the differential analyzer. Selecting a subgroup from each, Goldstine "assigned six of the best computers to learn how to program the ENIAC and report to [John] Holberton," employed by the Army Ordnance Department to supervise civilians.[41] With no precedents from either sex, the creation and gendering of "computer operator" offers insight into how sexual divisions of labor gather momentum. Computing was a female job, and other female clerical workers operated business machines. So it was not unusual that in July 1945, women would migrate to a similar but new occupation. The six women—Kathleen McNulty, Frances Bilas, Betty Jean Jennings, Ruth Lichterman, Elizabeth

Snyder, and Marlyn Wescoff—reported to the Moore School to learn to program the ENIAC.

The ENIAC project made a fundamental distinction between hardware and software: designing hardware was a man's job; programming was a woman's job. Each of these gendered parts of the project had its own clear status classification. Software, a secondary, clerical task, did not match the importance of constructing the ENIAC and getting it to work.[42] The female programmers carried out orders from male engineers and army officers. It was these engineers and officers, the theoreticians and managers, who received credit for invention. The U.S. Army's social caste system is historically based on European gentlemen's social codes.[43] As civil servants, the six women computers chosen to operate the ENIAC stood outside this system.

Yet if engineers originally conceived of the task of programming as merely clerical, it proved more complex. Under the direction of Herman and Adele Goldstine, the ENIAC operators studied the machine's circuitry, logic, physical structure, and operation. Kathleen McNulty described how their work overlapped with the construction of the ENIAC: "Somebody gave us a whole stack of blueprints, and these were the wiring diagrams for all the panels, and they said 'Here, figure out how the machine works and then figure out how to program it.' This was a little bit hard to do. So Dr. Burks at that time was one of the people assigned to explain to us how the various parts of the computer worked, how an accumulator worked. Well once you knew how an accumulator worked, you could pretty well be able to trace the other circuits for yourself and figure this thing out."[44]

Understanding the hardware was a process of learning by doing. By crawling around inside the massive frame, the women located burnt-out vacuum tubes, shorted connections, and other nonclerical bugs.[45] Betty Jean Jennings's description confirms the ingenuity required to program at the machine level and the kinds of tacit knowledge involved:

> We spent much of our time at APG learning how to wire the control board for the various punch card machines: tabulator, sorter, reader, reproducer, and punch. As part of our training, we took apart and attempted to fully understand a fourth-order difference board that the APG people had developed for the tabulator. . . . Occasionally, the six of us programmers all got together to discuss how we thought the machine worked. If this sounds haphazard, it was. The biggest advantage of learning the ENIAC from the diagrams was that we began to understand what it could and what it could not do. As a result we could diagnose troubles almost down to the individual vacuum tube. Since we knew both the application and the machine, we learned to diagnose troubles as well as, if not better than, the engineer.[46]

Framing the ENIAC story as a case study of the mechanization of female labor, it would be hard to argue that de-skilling accompanied mechanization.[47] The idiom of sex-typing, which justified assigning women to software, contradicted the actual job, which required sophisticated familiarity with hardware. The six ENIAC operators understood not only the mathematics of computing but the machine itself. That project leaders and historians did not value their technical knowledge fits the scholarly perception of a contradiction between the work actually performed by women and the way others evaluate that work. In the words of Nina Lerman, "Gender plays a role in defining which activities can readily be labeled 'technological.'"[48]

Meanwhile, at the Los Alamos Scientific Laboratory in New Mexico, scientists were preparing a new thermonuclear weapon, the Super. Stanley Frankel and Nicholas Metropolis, two Los Alamos physicists, were working on a mathematical model that might help to determine the possibility of a thermonuclear explosion. John Von Neumann, a technical consultant, suggested that Los Alamos use ENIAC to calculate the Super's feasibility. Once Von Neumann told Herman Goldstine about this possible use, Herman and Adele invited Frankel and Metropolis to Philadelphia and offered them training on the ENIAC. When the two physicists arrived in Philadelphia in the summer of 1945, Adele Goldstine and the women operators explained how to use the machine. McNulty recalled that "We had barely begun to think that we had enough knowledge of the machine to program a trajectory, when we were told that two people were coming from Los Alamos to put a problem on the machine."[49] Despite such self-effacing comments, the operators demonstrated impressive mastery of the ENIAC during the collaboration with the Los Alamos physicists. By October, the two theoretical physicists had programmed their elaborate problem on huge sheets of paper. Then, the women programmed it into the machine, which no one had formally tested. As McNulty explained, "No one knew how many bad joints there were, and how many bad tubes there were, and so on."[50] The cooperative endeavor furthered the operators' intimate understanding of ENIAC as they pushed it to a new level of performance. Programming for Frankel and Metropolis took one million IBM punch cards, and the machine's limited memory forced the women to print out intermediate results before repunching new cards and submitting them to the machine. Within a month, the Los Alamos scientists had their answer—that there were several design flaws.[51]

The "ENIAC girls" turned their attention back to shell trajectory calculations and were still engaged on that project when the war ended. The ENIAC, designed and constructed in military secrecy, was prepared for public unveiling in early 1946. A press conference on 1 February and a formal dedication on 15 February each featured demonstrations of the machine's capabilities. According to Herman Goldstine, "The actual preparation of the problems put on at the demonstration was done by Adele Goldstine and me with some help on the simpler problems from John Holberton and his girls."[52] Indeed, Elizabeth Snyder and Betty Jean Jennings developed the demonstration trajectory program.[53] Although women played a key role in preparing the demonstrations, both for the press and for visitors to the laboratory, this information does not appear in official accounts of what took place.

Contemporary Accounts of ENIAC

Social constructionist historians and sociologists of science take the position that scientists describing their experimental work do not characterize events as they actually happened.[54] Publicity for technical demonstrations is not so different. In presenting ENIAC to the public, engineers staged a well-rehearsed event. They cooperated with the War Department, which controlled representations of the project through frequent press releases to radio and newspapers.

It is a curious paradox that while the War Department urged women into military and civil service and fed the media uplifting stories about women's achievements during the war, its press releases about a critical project like the ENIAC do not mention the women who helped to make the machine run. War Department press releases characterize ENIAC as "designed and constructed for the Ordnance Department at the Moore School of Electrical Engineering of the University of Pennsylvania by a pioneering group of Moore School experts."[55] They list three individuals as "primarily responsible for the extremely difficult technical phases of work . . . Eckert—engineering and design; Mauchly—fundamental ideas, physics; Goldstine—mathematics, technical liaison."[56] The War Department's selective press releases highlighted certain individuals involved in the ENIAC project while omitting others, specifically the women operators. Because of these omissions the operators were neither interviewed nor offered the opportunity to participate in telling the ENIAC story. Newspaper accounts characterize ENIAC's ability to perform tasks as "intelligent" but the women doing the same computing tasks did not receive similar acclaim.[57] While the media publicly hailed hardware designers as having "fathered" the machine, they did not mention women's contributions. The difference in status between hardware and software illustrates another chapter in the story of women in the history of science and technology. The unmentioned computer technicians are reminiscent of Robert Boyle's "host of 'laborants,' 'operators,' 'assistants' and 'chemical servants'" whom Steven Shapin described as "invisible actors." Working three centuries earlier, their fate was the same: they "made the machines work, but they could not make knowledge."[58]

The *New York Times* of 15 February 1946 described Arthur Burks's public demonstration: "The ENIAC was then told to solve a difficult problem that would have required several weeks' work by a trained man. The ENIAC did it in exactly 15 seconds."[59] The "15 seconds" claim ignores the time women spent setting up each problem on the machine. Accompanying photographs of Eckert and Mauchly, the article reported that "the Eniac was invented and perfected by two young scientists of the [Moore] school, Dr. John William Mauchly, 38, a physicist and amateur meteorologist, and his associate, J. Presper Eckert Jr., 26, chief engineer on the project. Assistance was also given by many others at the school. . . . [The machine is] doing easily what had been done laboriously by many trained men. . . . Had it not been available the job would have kept busy 100 trained men for a whole year."[60] While this account alludes to the participation of many individuals other than Eckert and Mauchly, the hypothetical hundred are described as men. Why didn't the article report that the machine easily did calculations that would have kept one hundred trained women busy, since BRL and the Moore School hired women almost exclusively as computers? Even in an era when language defaulted to "he" in general descriptions, this omission is surprising, since the job of computer was widely regarded as women's work.[61] Women seem to have vanished from the ENIAC story, both in text and in photographs. One photograph accompanying the *New York Times* story foregrounds a man in uniform plugging wires into a machine. While the caption describes the "attendants preparing the machine to solve a hydrodynamical problem," the figures of two women in the background can be seen only by close scrutiny (Figure 11-1). Thus, the press conference and follow-up coverage rendered invisible both the skilled labor required to set up the demonstration and the gender of the skilled workers who did it.

The role of the War Department and media in shaping public discourse about the machine and its meaning is significant. Several potential opportunities for the women operators to get some public attention and credit for their work never materialized. For example, the publicity photograph of the ENIAC printed in the *New York Times* was among the most widely disseminated images of the machine. When it was published as an army recruitment advertisement (Figure 11-2), the women were cropped out.[62] This action is understandable, at one level, since the operators were all civilians. Yet given the important participation of WACs in closely related wartime work, it constituted another missed opportunity to give the women their due.

Figure 11-1. One of the most widely reprinted photographs of ENIAC.
U.S. Army Photo, http://ftp.arl.army.mil/ftp/historic-computers/png/eniac2.png

Archival records show that photographers came in to record the ENIAC and its engineers and operators at least twice. Neither visit resulted in any publicity for the women. On the first occasion, an anonymous photographer's pictures of the ENIAC group turned out poorly. Herman Goldstine wrote apologetically to Captain J. J. Power, Office of the Chief of Ordnance: "Dear John, I am returning herewith the photographs with sheets of suggested captions. As you can see from looking at these photographs, many of them are exceedingly poor, and, I think, unsuitable for publication."[63] Nonetheless, the captions for these unsuitable photographs are instructive:

> VIEW OF ONE SIDE OF THE ENIAC: Miss Frances Bilas (Philadelphia, Pa.) and Pfc. Homer W. Spence (Grand Rapids, Mich.) are setting program switches. Miss Bilas is an ENIAC operator in the employ of the Ballistic Research Laboratory, Aberdeen Proving Ground, Md., and Pfc. Spence is a maintenance engineer. . . .

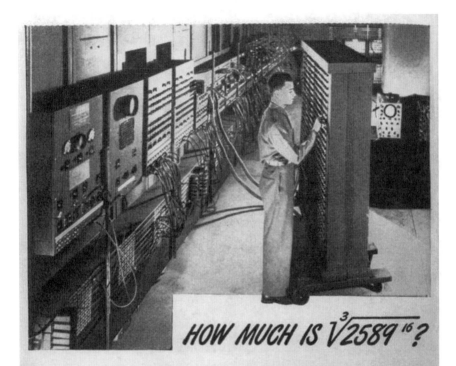

HOW MUCH IS $\sqrt[3]{2589}^{16}$?

The Army's ENIAC can give you the answer in a fraction of a second!

Think that's a stumper? You should see *some* of the ENIAC's problems! Brain twisters that if put to paper would run off this page and feet beyond . . . addition, subtraction, multiplication, division — square root, cube root, any root. Solved by an incredibly complex system of circuits operating 18,000 electronic tubes and tipping the scales at 30 tons!

The ENIAC is symbolic of many amazing Army devices with a brilliant future for you! The new Regular Army needs men with aptitude for scientific work, and as one of the first trained in the post-war era, you stand to get in on the ground floor of important jobs

YOUR REGULAR ARMY SERVES THE NATION AND MANKIND IN WAR AND PEACE

which have never before existed. You'll find that an Army career pays off.

The most attractive fields are filling quickly. Get into the swim while the getting's good! 1½, 2 and 3 year enlistments are open in the Regular Army to ambitious young men 18 to 34 (17 with parents' consent) who are otherwise qualified. If you enlist for 3 years, you may choose your own branch of the service, of those still open. Get full details at your nearest Army Recruiting Station.

A GOOD JOB FOR YOU
U. S. Army
CHOOSE THIS
FINE PROFESSION NOW!

Figure 11-2. This advertisement appeared in *Popular Science Monthly*, October 1946. Army materials courtesy of the U.S. Government, as represented by the Secretary of the Army.

SETTING UP A PROBLEM ON THE ENIAC: Reading from left to right, Miss Akrevoe Kondopria (Philadelphia, Pa.) at an accumulator, Miss Betty Jennings (Stanbury, Mo.), Cpl. Irwin Goldstein (Brooklyn, NY) and Miss Ruth Lichterman (Rockaway, NY) standing at function tables. Miss Kondopria is a Moore School employee on the ENIAC project; Miss Jennings and Miss Lichterman are ENIAC operators employed by the Ballistic Research Laboratory, Aberdeen Proving Ground, Md., and Cpl. Goldstein is a maintenance engineer. . . .

SETTING UP A PROBLEM ON THE ENIAC: Reading from left to right, Miss Betty Snyder (Narberth, Pa.), Miss Betty Jennings (Stanbury, Mo.), Miss Marlyn Wescoff (Philadelphia, Pa.) and Miss Ruth Lichterman (Rockaway, NY). Miss Snyder is setting program switches on an accumulator; Miss Jennings is setting up numbers to be remembered in the function table . . . Miss Wescoff and Miss Lichterman are working at the printer. . . . The function table which stores numerical data set up on its switches is seen at the right and its two control panels are behind Miss Frances Bilas (Philadelphia, Pa.) who is plugging a program cable in the master programmer. Miss Bilas is an ENIAC operator in the employ of the Ballistic Research Laboratory, Aberdeen Proving Ground, Maryland.[64]

"Setting switches," "plugging cables," and "standing at function tables"—such captions understate the complexities of women's work. While two men appear alongside the operators, they are "maintenance engineers," occupational titles suggesting technical expertise.

The second photographer was Horace K. Woodward Jr., who wrote an article about ENIAC for *Science*. He wrote to Adele Goldstine: "Dear Mrs. Goldstine and other mENIACS, You will be perturbed to hear that the color flesh shots (oops, flAsh shots) that I was taking 1 Feb 46 turned out nicely. I hadn't intended them for publication but thought you folks might like them."[65] His article in *Science* carried no photographs of the women and made no reference to their existence.

More surprising still, the media reports did not highlight Adele Goldstine, despite her leadership position and her expertise in a technical realm that had not earlier existed for either sex.[66] An affidavit Adele Goldstine submitted as testimony in Sperry Rand v. Bell Labs explains how she saw her own role: "I did much of the programming and the setting up of the ENIAC for the various problems performed on it while I was at the Moore School. I also assisted my husband in training Mr. Holberton and a group of girls to set up problems on the ENIAC. . . . I worked with Mr. Holberton and his group to program each problem which they put on the ENIAC up to and including the demonstration problems for the ENIAC dedication exercises."[67] Adele Goldstine and Moore School professor Harry Huskey were charged with producing an ENIAC operating manual, a complete technical report, and a maintenance manual.[68] Herman Goldstine explains: "The only persons who really had a completely detailed knowledge of how to program the ENIAC were my wife and me. Indeed, Adele Goldstine wrote the only manual on the operation of the machine. This book was the only thing available which contained all the material necessary to know how to program the ENIAC and indeed was its purpose."[69] In addition, he reports that his wife contributed heavily to a 1947 paper he coauthored with John Von Neumann, "Planning and Coding Problems for an Electronic Computing Instrument".[70]

It is an overstatement to say that female computers and operators were never covered in any media. A few articles mention them, as in this example:

> An initial group, consisting primarily of women college graduates, especially trained for work by the Moore School, began the work in ground gunfire, bombing and related ballistics studies immediately after Pearl Harbor, when the Aberdeen Proving Ground's Ballistic Research Laboratory broadened its program at the University.

> Forerunners of a group eventually numbering more than 100, they made use of the Moore School's differential analyzer, which is equally useful in the realm of ballistics and the solution of peacetime mathematical problems.

> Two other groups were organized later, under separate contracts, one of which was devoted to analysis of experimental rocket firing at Aberdeen, while the other assisted in the proving ground development of new shells and bombs.[71]

This recognition is quite different from the publicity accorded to male officers and engineers associated with the project.[72] The article cited here portrays the women as interchangeable. Even if it were too space-consuming to name each human computer, it is still notable that the article describes the women as being trained for work "by the Moore School" as opposed to "by Adele Goldstine" or by her many female colleagues.[73] That ENIAC's 1946 demonstration doubled as a vanishing act for its female participants fits neatly with postwar propaganda that as early as 1944 began redirecting women into more traditional female occupations or out of the paid labor force entirely.[74]

And what of the several years after World War II? While the Department of Labor acknowledged women's desire to stay on in paid employment, its publications were not so optimistic.[75] An avalanche of materials urged women to leave work. A 1948 *Women's Bureau Bulletin* reported on the situation for women with mathematics education who sought paid work:

> Although, during the war, production firms and Government projects were important outlets for women trained in mathematics, the emphasis, following the end of hostilities, shifted back to the more usual channels. Teaching and employment with insurance and other business firms became the principal outlets for women college graduates with mathematical training. . . . Most of the wartime research projects sponsored by the Government were dropped after V-J day. In the few that continued, the small number of mathematical jobs were filled by the staffs of the institutions at which the research was being done and by men with mathematical skills who were being released from military service. The women's military services, which utilized women with mathematical training during the war, were reduced to very small staffs. . . . As women leave, men will be hired to replace them. . . . Although many women are continuing on their wartime mathematical jobs, it is difficult to say how much of the gain will be in terms of permanent opportunities for women.[76]

The Federal Bureau of Investigation dropped many of the women it had hired as cryptographers during the war. By 1946, the National Bureau of Standards had filled most of the vacancies on its computing staff with male veterans.[77] At the Ballistics Research Laboratory, an army memorandum detailed criteria for how individuals would be let go, with separate instructions for male officers and for WAC officers.[78] With this in mind, the absence of women from an October 1946 army recruitment ad makes sense. The "propaganda machine," as Herzenberg and Howes call it, that during the war had so successfully called women out of their homes, made a 180-degree turn, pushing many women back towards full-time domesticity.[79]

In the 1950s, new opportunities developed alongside continuing ambivalences about women's occupational roles. A 1956 U.S. Department of Labor report on employment opportunities for women mathematicians and statisticians is replete with examples of women's mathematical work—and the future need for women mathematicians—in a variety of fields including programming. Four "findings" appear as an executive summary:

1. More women mathematicians and statisticians are currently needed, and interesting jobs await those trained at the bachelor's degree as well as graduate levels.
2. Young women in high school should be encouraged to try mathematics and if they have the qualifications for success in mathematics and statistics should be encouraged to prepare for those fields; anticipated shortages make the long-run outlook exceptionally favorable.
3. Young women who combine the qualifications for teaching with ability in mathematics should be encouraged to teach, at least part time, since in teaching they can magnify their contribution to the Nation's progress.
4. Mature college women who have majored in mathematics, possess the personal qualifications for teaching, and have time available to work, should prepare themselves through refresher courses in mathematics and education for teaching positions, if they live in one of the many communities experiencing or anticipating a shortage of mathematics teachers.[80]

The report explores a wide range of career options, including programming and actuarial work. Yet as the patriotic rhetoric of service "to the Nation's progress" makes clear, the Department of Labor prioritized teaching as a career choice. Science and engineering had won the war, and now the developing baby boom predicted a growing demand for math teachers.

Despite such exhortations, some women never left computer programming. Fran Bilas, Kay McNulty, and Betty Snyder continued briefly with ENIAC when it moved to BRL in 1947; Ruth Lichterman stayed on for two years.[81] Other women joined the ENIAC at BRL following the war. Betty Snyder Holberton went on to program UNIVAC and to write the first major software routine ever developed for automatic programming. She also collaborated on writing COBOL and FORTRAN with Grace Hopper, a key programmer of the Mark I. Hopper left active duty with the U.S. Navy as a lieutenant in 1946 but remained with the Navy reserves until 1966. From 1946 until it started running programs around 1951, the Electronic Computer Project at Princeton's Institute for Advanced Study employed mostly female programmers, who included Thelma Estrin, Hedi Selberg, Sonia Bargmann, and Margaret Lamb. Their accomplishments are future chapters for a history of computer programming.

Conclusion

The ENIAC story highlights several issues in the history and historiography of gender, technology, and labor. Major wars have unmistakable influences on gender relations and work, and those effects can be elusive and complex. Conflicts among representations of women's work in computing ensure work for the historian in distinguishing seeming gender changes from real ones. These conflicts and sometime contradictions lie at the heart of women's historical invisibility.

First, the variance between effusive wartime recruiting literature and historians' evaluations of women's actual opportunities is striking. Disputing the claims of propaganda, historians generally agree that during wartime women may have made some progress in expanding the varieties of work they could do. Yet rather than move up the ladder of success women's work appears to have added more rungs at the bottom. The narrative histories of the ENIAC since 1946 echo this finding. With few exceptions, they make the implicit or explicit assumption that, while women were involved, their participation was not sufficiently important to merit explication. Thus, this episode in the history and historiography of computing confirms Rossiter's "Matilda effect": individuals at the top of professional hierarchies receive repeated publicity and become part of historical records, while subordinates do not, and quickly drop from historical memory.[82]

A second conflicting representation concerns the actual work performed by women contrasted with how employers categorized this work. As this article shows, the evidence of ENIAC challenges the implicit assumption of computing historians that the low-status occupations of women meant that their work could not be innovative. Wartime propaganda proclaimed "no limitations on your opportunities," yet only certain jobs were open to women. However, it was within the confines of precisely such low-status occupational classifications that women engaged in unprecedented work. Looking behind media accounts and later narratives of the development of ENIAC to consider primary source accounts of the work women actually performed reveals how its low-status categorization clashed with the kinds of knowledge required. Finding this mismatch offers the possibility that, in their work as operators, women moving into stereotypical male domains played a subversive role, challenging the gender status quo before the war. According to this view, women's invisibility reflects deep-rooted ambivalences about the roles women professionals began to occupy in the labor force. These ambivalences permeated both power relationships in the workplace and media portrayals of women's contributions.

Third, portrayals of women's postwar fate continue the ambivalence that characterized their wartime work. Women were seen as meeting a crisis—but only a temporary one. One 1943 guide to managers explained: "Women can be trained to do any job you've got—but remember 'a woman is not a man;' A woman is a substitute—like plastic instead of metal."[83] Both postwar propaganda and historians characterize women as retreating to teaching and homemaking after the war, abandoning their gains. Yet a fair number did not leave the workforce, a fact that the Department of Labor acknowledged even as it urged women toward teaching.[84]

The revised history of ENIAC presented here reveals that many of historians' questions about the history of computing reflect the unintentionally "male-centered terms" of history.[85] The result is a distorted history of technological development that has rendered women's contributions invisible and promoted a diminished view of women's capabilities in this field. These incomplete stories emphasize the notion that programming and coding are, and were, masculine activities. As computers saturate daily life, it becomes critical to write women back into the history they were always a part of, in action if not in memory.

Notes

1. History has valued hardware over programming to such an extent that even the *IEEE Annals of the History of Computing* issue devoted to ENIAC's fiftieth anniversary barely mentioned these women's roles. See *IEEE Annals of the History of Computing* 18, no. 1 (1996). Instead, they were featured two issues later in a special issue on women in computing.
2. Ruth Milkman, *Gender at Work: The Dynamics of Job Segregation by Sex During World War II* (Chicago, 1987).
3. Two books currently offer some information on the participation of women in computer history: see Autumn Stanley, *Mothers and Daughters of Invention: Notes for a Revised History of Technology* (Metuchen, N.J., 1993), and Herman Goldstine, *The Computer from Pascal to Von Neumann* (Princeton, 1972). For recollections from women who worked on the ENIAC, see W. Barkley Fritz, "The Women of ENIAC," *IEEE Annals of the History of Computing* 18, no. 3 (1996): 13–28. Other histories tend to make passing references to the women and to show photographs of them without identifying them by name.
4. Evelyn Steele, *Wartime Opportunities for Women* (New York, 1943), preface. For an analysis of American mobilization propaganda directed at women, see Leila Rupp, *Mobilizing Women for War: German and American Propaganda, 1939–1945* (Princeton, 1978).
5. Keith Ayling, *Calling All Women* (New York, 1942), 129.

6. Steele, 101.

7. Ibid., 99–100.

8. According to a *Women's Bureau Bulletin*, "A coeducational university, which before the war had few outlets for mathematics majors except in routine calculating jobs, found many attractive jobs available to mathematics majors during the war, mostly in Government-sponsored research. . . . There was a definite shift from the usual type of employment for mathematics majors in teaching and in clerical jobs in business firms to computing work in industry and on Government war projects." See United States Department of Labor, "The Outlook for Women in Mathematics and Statistics," *Women's Bureau Bulletin* 223–24 (1948): 3. According to this report, women comprised the majority of high-school mathematics teachers.

9. Ibid., 8. Margaret Rossiter, *Women Scientists in America: Before Affirmative Action, 1940–1972* (Baltimore, 1995), 13, confirms this practice more widely in the sciences. The few women who worked in supervisory roles generally supervised other women, a much less prestigious managerial role than supervising men. However, at the Work Project Administration's Mathematical Tables Project, women supervised male computers. See Denise W. Gürer, "Women's Contributions to Early Computing at the National Bureau of Standards," *IEEE Annals of the History of Computing* 18, no. 3 (1996): 29–35. The War Department in 1942 classified all military occupational specialties as either suitable or unsuitable for women; all jobs involving supervision over men were automatically declared unsuitable. Public Law 110 also made explicit that women could not command men without intervention from the secretary of war; see Bettie Morden, *The Women's Army Corps, 1945–1978* (Washington, D.C., 1990), 14.

10. See Margaret Rossiter, *Women Scientists in America: Struggles and Strategies to 1940* (Baltimore, 1982), also *Women Scientists in America: Before Affirmative Action, 1940–1972*. In the 1982 volume, p. 55, Rossiter describes the late-nineteenth-century star counters in astronomical laboratories who performed computer work for male astronomers. The famed astronomer Maria Mitchell was employed as a computer for the U.S. Coast and Geodetic Survey in the late 1860s. The term computer, meaning "one who computes," originally referred to the human who was assigned various mathematical calculations. Ute Hoffman dates the use of computer to the seventeenth century, when it was used in reference to men who tracked the course of time in their calendars. For decades the terms computer and calculator were interchangeable. In fact, early computers such as the ENIAC and Mark I were called electronic calculators. See Ute Hoffmann, "Opfer und Täterinnen: Frauen in der Computergeschichte," in *Micro Sisters: Digitalisierung des Alltags, Frauen und Computer*, ed. Ingrid Schöll and Ina Küller (Berlin, 1988). A number of other historians have documented women's work in other sciences. For example, Peter Galison, *Image and Logic: A Material Culture of Microphysics* (Chicago, 1997), discusses the work of women in high-energy physics laboratories, both those who counted flashes on the scintillator in Rutherford's laboratory and those who scanned the photographs from

bubble-chamber experiments. Caroline Herzenberg and Ruth Howes, "Women of the Manhattan Project," *Technology Review* 8 (1993): 37, describe the work of women at Los Alamos, "some with degrees in mathematics and others with little technical background," who performed mathematical calculations for the design of the bomb. Amy Sue Bix, "Experiences and Voices of Eugenics Field-Workers: 'Women's Work' in Biology," *Social Studies of Science* 27 (1997): 625–68, reports the work of female field-workers at the Eugenics Record Office, who gathered data on individuals and families. In every case the work was subordinate to men's. See also Jane S. Wilson and Charlotte Serber, eds., *Standing By and Making Do: Women of Wartime Los Alamos* (Los Alamos, N.M., 1988).

11. See Rossiter, *Women Scientists in America* (both volumes). According to Herman Goldstine, it was the fact that women were not seeking career advancement that made them ideal workers: "In general women didn't get Ph.D.'s. You got awfully good women because they weren't breaking their backs to be smarter than the next guy." Herman Goldstine, interview by author, Philadelphia, 16 November 1994. Goldstine also noted that the few men he encountered working on programming rarely conceived of their jobs as permanent. Rather, they were steps on the way to something better. These jobs were "never careers for them, but a way of making money for a short time." Consequently, Goldstine observes, "Men in general were lousy—the brighter the man the less likely he was to be a good programmer. . . . The men we employed were almost all men who wanted Ph.D.'s in math or physics. This [hands-on work] was a bit distasteful. I think they viewed what they were doing as something they were not going to be doing for a career. If you take a woman like Hedi Selberg [a programmer at the Institute for Advanced Study Electronic Computer Project] she probably didn't want to sit around with the baby all the time."

12. Galison cites the invention and popularization of the term "scanner girl."

13. Ibid., 176.

14. For further discussion of prewar trends in hiring practices, see Lisa Fine, *The Souls of the Skyscraper: Female Clerical Workers in Chicago, 1870–1930* (Philadelphia, 1990), and Margery Davies, *Women's Place is at the Typewriter: Office Work and Office Workers, 1870–1930* (Philadelphia, 1982). See also Milkman (n. 2 above), chaps. 1–3.

15. Paul Ceruzzi, "When Computers Were Human," *Annals of the History of Computing* 13 (1991): 242.

16. Cf. Milkman, 49: "The boundaries between 'women's' and 'men's' work changed location, rather than being eliminated. . . . Rather than hiring women workers to fill openings as vacancies occurred, managers explicitly defined some war jobs as 'suitable' for women, and others as 'unsuitable,' guided by a hastily revised idiom of sex-typing that adapted prewar traditions to the special demands of the war emergency." Both Milkman and Fine discuss how gender-specific advertisements reflect the feminization of specific occupations. Fine offers an analysis of the shifting gender imagery of some clerical occupations. On this point, however, note that focusing on the industry's language about women (in this case, the stories about the biological capacities and natural implications of

womanhood—or, by extension, on the advertising techniques used to create a gendered labor force) can confuse industry ideals with women's actual practice. As Milkman's notion of the idiom of sex-typing suggests, there is indeed a disjuncture between women's prescribed place and what women actually did. This disjuncture is central to women's invisibility in technological history.

17. Goldstine interview (n. 11 above). The domain's masculinity appears in the preface of a textbook on exterior ballistics: Office of the Chief of Ordnance, *The Method of Numerical Integration in Exterior Ballistics: Ordnance Textbook* (Washington, D.C., 1921). "The names of the men who have contributed most to its [the text's] development, particularly Major Moulton and Professor Bliss, are mentioned in various places in the text, and to whom the writer might appropriately make personal acknowledgement, would amount practically to an enumeration of all the officers, civilian investigators, and computers who have been connected with the work in ballistics in Washington and at the Aberdeen Proving Ground."

18. The heads of the computing groups were all college graduates, as were the majority of computers.

19. "The title 'engineering computer' was created for these women, since such work before the war was done by young, junior engineers as part of their induction training following graduation from an engineering college." U.S. Department of Labor, "Women in Architecture and Engineering," *Women's Bureau Bulletin* 223–25 (1948): 56. See Sharon Hartmann Strom, *Beyond the Typewriter: Gender, Class, and the Origins of Modern American Office Work, 1900–1930* (Urbana, Ill., 1992), for a discussion of similar circumstances within American businesses. To call a particular job "feminized" does not restrict it to women. Certainly there were some male computers and programmers. For a review of literature on gender and technology, see Nina Lerman, Arwen Palmer Mohun, and Ruth Oldenziel, "Versatile Tools: Gender Analysis and the History of Technology," *Technology and Culture* 38 (1997): 1–30.

20. The idiom of sex-typing made the sexual division of labor seem natural; differences in work capacity were considered biologically based. Evelyn Steele, editorial director of Vocational Guidance Research, writes, "It is generally agreed that women do well at painstaking, tedious work requiring patience and dexterity of the hands. The actual fact that women's fingers are more slender than men's makes a difference. Also, women adapt themselves to repetitive jobs requiring constant alertness, nimble fingers and tireless wrists. They have the ability to work to precise tolerances, can detect variations of ten-thousandths of an inch, [and] can make careful adjustments at high speed with great accuracy"; Steele (n. 4 above), 46. Women's strengths thus lay in performing repetitive, detailed, unskilled tasks. Such statements were not new. Arguments made in favor of women working as telephone operators were similar: "The work of successful telephone operating demanded just that particular dexterity, patience and forbearance possessed by the average woman in a degree superior to that of the opposite sex." Brenda Maddox, "Women and the Switchboard," in *The Social Impact of the Telephone*, ed. Ithiel de Sola Pool (Cambridge, Mass., 1977), 266.

See also Fine (n. 14 above), chap. 4, "The Discourse on Fitness: Science and Symbols." For a discussion of women's wartime labor as portrayed in literature and advertising, see Charles Hannon, " 'The Ballad of the Sad Cafe' and Other Stories of Women's Wartime Labor," *Genders* 23 (1996): 97–119.

21. For a further discussion of the prewar situation and the complex interaction between new technologies and the sexual division of labor, see Fine, also Davies (n. 14 above). Jobs with a more established tradition of male employment were less likely to become feminized before World War II. For example, while "clerk" and "bookkeeper" stayed largely male, feminization was more widespread in stenography because it had not been defined as male. See Milkman (n. 2 above), chap. 4. For further discussion of how new jobs were gendered, see Heidi Hartmann, Robert Kraut, and Louise Tilly, eds., *Computer Chips and Paper Clips: Technology and Women's Employment*, 2 vols. (Washington, D.C., 1986), vol. 1, chap. 2.

22. See Rossiter, *Women Scientists in America: Struggles and Strategies to 1940* (n. 10 above), and Milkman.

23. At the time, women were concentrated in clerical roles more than in any other occupation; they comprised 54 percent of all clerical workers in 1940 and 62 percent in 1950. U.S. Department of Labor, "Changes in Women's Occupations 1940–1950," *Women's Bureau Bulletin* 253 (1954): 37. Clerical work encompasses a broad range of jobs, including office machine operators. The Employment and Training Administration and U.S. Employment Service's *Dictionary of Occupational Titles* (Washington, D.C., 1939–41) classified computing-machine operator and calculating-machine operator as entry-level clerical occupations. For further discussion of the wide range of clerical jobs, see Strom (n. 19 above) and Fine. See also David Alan Grier, "The ENIAC, the Verb 'to program' and the Emergence of Digital Computers," *IEEE Annals of the History of Computing* 18, no. 1 (1996): 51–55.

24. It was part of a prior agreement with the Moore School that in times of national emergency the Aberdeen Proving Ground could commandeer the school's differential analyzer. Lydia Messer, oral history, interview by Cornelius Weygandt, 22 March 1988, University of Pennsylvania Archives, Philadelphia. Joel Shurkin, *Engines of the Mind* (New York, 1984), 119. BRL had apparently organized previous cooperative projects during World War I with the University of Pennsylvania. The U.S. Army Ordnance Department's *Course in Exterior Ballistics: Ordnance Textbook* (Washington, D.C., 1921) credits H. H. Mitchell of the University of Pennsylvania as "Master Computer, who organized the range table computation work at Aberdeen." Before 1941, the Moore School also provided computers for BRL. Nancy Stern, *From ENIAC to UNIVAC: An Appraisal of the Eckert-Mauchly Computers* (Bedford, Mass., 1981), 10.

25. Shurkin, 128.

26. Shurkin, 127–28.

27. Stern, 13–14.

28. Not all women's jobs ranked lower or earned less than men's, but the history of female employment shows a persistent pattern into which the BRL's policies

fit. For example, see Sharon Hartmann Strom, "'Machines Instead of Clerks': Technology and the Feminization of Bookkeeping, 1910–1950," in Hartmann, Kraut, and Tilly (n. 21 above), 2:63–97. See Fritz (n. 3 above) for women's accounts of the work they performed and H. Polachek, "Before the ENIAC," *IEEE Annals of the History of Computing* 19, no. 2 (1997): 25–30, for the complexities of computations for preparing firing tables.

29. Adele Goldstine received her bachelor's degree from Hunter College in 1941, then a master's from the University of Michigan in 1942. In 1942 she taught mathematics in the public school system in Philadelphia. From late 1943 to March 1946 she worked for the ENIAC project at the Moore School and spent part of 1944 at the Aberdeen Proving Ground. In 1948, she resumed graduate study at New York University. She became a consultant to the Atomic Energy Commission project effective 7 June 1947, working on making the ENIAC into a stored-program computer. Herman Goldstine recalls that "Los Alamos was the major user of the ENIAC so it was [John] Von Neumann [who was using it]. Adele was his assistant. I was also a consultant but she was doing the major part." Goldstine interview (n. 11 above).

30. Ibid.

31. Harold Pender to George McCelland, 23 July 1943, Information Files: World War II: WAC Training: Miscellaneous, University of Pennsylvania Archives.

32. *Daily Pennsylvanian*, 29 September 1943, untitled clipping in Information Files: World War II: WAC Training: Miscellaneous, University of Pennsylvania Archives. While women received instructions from civilians (not an unusual practice in the armed services), they were commanded by military second lieutenants and corporals. The WAC officer in charge of the detachment on campus was Lt. Mildred Fleming.

33. Mattie Treadwell, *United States Army in World War Two Special Series: The Women's Army Corps* (Washington, D.C., 1954), 221.

34. Adele Goldstine to J. G. Brainerd, n.d., "Monday Night," Information Files: World War II: WAC Training: Miscellaneous, University of Pennsylvania Archives. The ES&MWTesses were the women involved in the Engineering, Science, and Management War Training courses. J. G. Brainerd was a professor at the Moore School and liaison with U.S. Army Ordnance.

35. Helen Rogan, *Mixed Company: Women in the Modern Army* (New York, 1981), 41; Treadwell, chap. 4. Building on the work of historians such as Milkman (n. 2 above) and Fine (n. 14 above), who have analyzed the need for women in men's jobs to maintain femininity, Leisa Meyer has described the sexual politics of women's entrance into military service; see "Creating G.I. Jane: The Regulation of Sexuality and Sexual Behavior in the Women's Army Corps During World War Two," *Feminist Studies* 18 (1992) 581–601, and *Creating G.I. Jane: Sexuality and Power in the Women's Army Corps during World War Two* (New York, 1996).

36. "Topics Included in the Engineering, Science, and Management War Training Courses for Members of the W.A.C. from Aberdeen Proving Ground," Information Files: World War II: WAC Training: Miscellaneous, University of

Pennsylvania Archives. There was a second training course in 1945; Herman Goldstine Papers, American Philosophical Society Library, Philadelphia (hereinafter Goldstine Papers).

37. Goldstine, *The Computer from Pascal to Von Neumann* (n. 3 above), 134; Fritz (n. 2 above). The histories of other sciences, in both Britain and the United States, show scientists' wives filling a number of the more senior women's positions in science. For example, Cecil Powell's wife Isobel led the scanning girls in Powell's laboratory, and Janet Landis Alvarez, wife of Luis Alvarez, trained the women bubble-chamber scanners at Berkeley. Among the computers at NACA were a number of engineers' wives. At the Los Alamos Scientific Laboratory, John Von Neumann's second wife, Klara Dan Von Neumann, became a programmer and helped to program and code some of the largest programs of the 1950s. Also at Los Alamos were Kay Manley, wife of John Manley, and Mici Teller, wife of Edward Teller, who performed mathematical calculations for the design of the bomb. For further discussion of couples in the sciences, see Helena M. Pycior, Nancy G. Slack, and Pnina G. Abir-Am, eds., *Creative Couples in the Sciences* (New Brunswick, N.J., 1996). According to Fritz, at least four computers married engineers at the Moore School after 1946. Frances Bilas married Homer Spence, Kathleen McNulty became Mauchly's second wife, and Elizabeth Snyder married John W. Holberton. According to Goldstine, Betty Jean Jennings (Bartik) married a Moore School engineer. Also at the Moore School were Eckert's first wife, a draftsman for the ENIAC project; Alice Burks, whose husband Arthur worked with Eckert and Mauchly on the ENIAC design; and Emma Lehmer, wife of Derrick Henry Lehmer, a computer and table compiler.

38. "Thanks for the Memory," presumably written by WACs at the Moore School, ca. 1943–44, Goldstine Papers.

39. In a retrospective analysis, Goldstine framed the computers' job as a prime candidate for mechanization due to its low skill: "Computing is thus subhuman in that it calls on very few of man's manifold abilities and yet is fundamental to many of his other activities, as Leibnitz so clearly perceived. This then is basically why computing was chosen as a human task to be mechanized"; Goldstine, *The Computer from Pascal to Von Neumann*, 343.

40. It is unclear exactly when this shift occurred. It was at least as early as February 1945, when George Stibbitz wrote in a report on relay computers for the National Defense Research Committee: "Human agents will be referred to as 'operators' to distinguish them from 'computers' (machines)." Ceruzzi (n. 15 above), 240.

41. Goldstine interview (n. 11 above). Interestingly, Milkman (n. 2 above) has discussed how jobs perceived as feminine in some places were quintessentially masculine in others—often within the same industry. The idiom of sex-typing, while consistent in individual factories, often differed among factories manufacturing the same product. On the Mark II computer at the Navy's Dahlgren Proving Ground, for instance, operators were male. This area deserves further study.

42. The terms hard and soft, as used to describe gendered tasks, are significant. For the hard and soft sciences, hard mastery and soft mastery are binary distinctions

in science and technology implying that the "hard" ways of knowing are men's domain; "soft" ways of knowing are more feminine. Goldstine, when interviewed, reported that he had resisted "there being a distinction" between hardware and software. He observed: "At the beginning, the hardware was the important thing, but as soon as you get beyond the bottleneck of making the computer," programming software became a new bottleneck. "They've automated the bejeezus out of making chips but not software." Ironically, by the time the process of making hardware was automated programming software had become a man's job and acquired higher status than it had had in the 1940s. See, for example, Phillip Kraft, "The Routinization of Computer Programming," *Sociology of Work and Occupations* 6 (1977): 139–55.

43. Jeanne Holm, *Women in the Military: An Unfinished Revolution*, rev. ed. (Novato, Calif., 1992), 73. Social mores, as well as a variety of rules and regulations, meant that women's qualifications had to surpass men's before they could compete for higher-level jobs within academia (including government-sponsored research) and industry. The army had higher selection criteria for female officers and enlisted personnel "than those for men in the same service" (p. 50). P.L. 110, the legislation converting the WAC to full military status, specified that "its commanding officer could never be promoted above the rank of colonel and its other officers above the rank of lieutenant colonel; its officers could never command men unless specifically ordered to do so by Army superiors" (Treadwell [n. 33 above], 220). Additionally, the War Department in 1943 set the ratio of female officers to enlisted women at one to twenty. Comparable figures for men were one to ten. Using the excuse of a surplus of male officers, it capped WAC officers by limiting entrants to the WAC Officer Candidate School but did not impose a similar limitation on male officers. None of the six women ENIAC operators held high status in academia or the military. Men at the Moore School who were not affiliated with the army, such as Harry Huskey or Arthur Burks, had visible academic appointments. See Rossiter, *Women Scientists in America: Before Affirmative Action, 1940–1972* (n. 9 above), for more on hierarchies, promotions, and payment in science.

44. Shurkin (n. 24 above), 188.

45. Kraft (n. 42 above), 141.

46. Fritz (n. 2 above), 19–20.

47. A number of historians have disputed de-skilling assumptions. For example, Sharon Hartmann Strom, "'Machines Instead of Clerks'" (n. 28 above), 64, describes in the case of bookkeeping machine operators how "workers continued to apply hidden skills of judgement and to integrate a number of tasks, particularly to jobs in the middle levels of bookkeeping, even though these jobs required the use of machines." Fine (n. 14 above), 84, claims that the stenographer-typist's job was more challenging than the copyist's whom she replaced. For a review of literature on gender, mechanization, and de-skilling, see Nina Lerman, Arwen Palmer Mohum, and Ruth Oldenziel, "The Shoulders We Stand On and the View from Here: Historiography and Directions for Research," *Technology and Culture* 38 (1997): 9–30. See also Kenneth Lipartito, "When

Women Were Switches: Technology, Work, and Gender in the Telephone Industry, 1890–1920," *American Historical Review* 99 (1994): 1075–111.

48. Nina Lerman, "'Preparing for the Duties and Practical Business of Life': Technological Knowledge and Social Structure in Mid-19th-Century Philadelphia," *Technology and Culture* 38 (1997): 36. Judy Wajcman, *Feminism Confronts Technology* (University Park, Penn., 1991), 37, observes: "Definitions of skill can have more to do with ideological and social constructions than with technical competencies which are possessed by men and not by women."

49. Shurkin, 188.

50. Ibid., 189.

51. C. Dianne Martin, "ENIAC: Press Conference That Shook the World," *IEEE Technology and Society Magazine* 14, no. 4 (1995): 3–10. Because the problem was classified, the equations remained concealed.

52. Goldstine, *The Computer from Pascal to Von Neumann* (n. 3 above), 229. For details of the kinds of calculations performed using ENIAC, see Arthur W. Burks and Alice R. Burks, "The ENIAC: First General-Purpose Electronic Computer," *Annals of the History of Computing* 3 (1981): 310–89. The Burks were another significant husband and wife team, publishing their story together; Alice R. Burks and Arthur W. Burks, *The First Electronic Computer: The Atanasoff Story* (Ann Arbor, Mich., 1988).

53. Fritz (n. 2 above), 20–21. Goldstine recalled bringing Douglas Hartree, a physicist who had built a differential analyzer in Britain, to the United States for a visit. "I got Kay McNulty to be his programmer and she was good and intelligent. The girls soon branched off independently and it was during that period that my wife was making ENIAC into a stored program computer"; Goldstine interview (n. 11 above).

54. See, for example, Bruno Latour, *Science in Action* (Cambridge, 1987).

55. U.S. War Department, Bureau of Public Relations, "Ordnance Department Develops All-Electronic Calculating Machines," press release, February 1946, Goldstine Papers.

56. U.S. War Department, Bureau of Public Relations, "History of Development of Computing Devices," press release, 15–16 February 1946, Goldstine Papers.

57. For media characterizations of ENIAC, see C. Dianne Martin, "The Myth of the Awesome Thinking Machine," *Communications of the ACM* 36, no. 4 (1993): 125, 127; see also Martin, "ENIAC" (n. 51 above), 3–10. Like the laundry industry that made its employees invisible by publicizing the tireless machines, the ENIAC was portrayed as doing almost all of the work; Arwen Mohun, "Laundrymen Construct their World: Gender and the Transformation of a Domestic Task to an Industrial Process," *Technology and Culture* 38 (1997): 97–120.

58. Steven Shapin, "The House of Experiment in Seventeenth-Century England," *Isis* 79 (1988): 395.

59. T. R. Kennedy, "Electronic Computer Flashes Answers, May Speed Engineering," *New York Times*, 15 February 1946.

60. Ibid.

61. The NACA memorandum (n. 15 above) specifically used she to describe the computers in its service. Women played salient roles in the demonstration of many domestic and business technologies, from sewing machines to typewriters to IBM office products, making their omission here all the more pointed.

62. See, for example, *Popular Science Monthly*, October 1946, 212.

63. Herman Goldstine to Captain J. J. Power, Office of the Chief of Ordnance, 17 January 1946, Goldstine Papers.

64. ENIAC file appended to Goldstine to Power, 17 January 1946.

65. Horace K. Woodward Jr. to Adele Goldstine, 23 February 1946, Goldstine Papers.

66. While Adele Goldstine did not receive media acknowledgement, she clearly had some status among her colleagues at the Moore School as the only woman working on the machine's hardware. Initially, she oversaw Holberton. As head of the WAC course, despite her civilian status, she had frequent contact with top administrators at both the Moore School and the Aberdeen Proving Ground. In a publicity folder, biographical profiles on approximately a dozen staff members at the Moore School connected with the ENIAC include J. Presper Eckert, John W. Mauchly, Herman H. Goldstine, John G. Brainerd, Arthur Burks, Harry Huskey, Cpl. Irwin Goldstein, and Pfc. Spence. Adele Goldstine is the only woman included.

67. The affidavit is included in a letter from Harry Pugh, at Fish, Richardson, and Neave, to Herman Goldstine, 12 December 1961, Goldstine Papers.

68. Goldstine, *The Computer from Pascal to Von Neumann* (n. 3 above), 200.

69. Ibid., 330.

70. Ibid., 255 n. 4.

71. "Studies at Penn Aided Artillery," undated clipping from unidentified newspaper, ENIAC Publicity Folder, Goldstine Papers.

72. See, for example, Allen Rose, "Lightning Strikes Mathematics," *Popular Science Monthly*, April 1946, 85, photo caption: "T. K. Sharpless, of the Moore School of Engineering, sets a dial on the Eniac's initiating unit, which contains some of the master controls of the huge, complex mechanics. . . . Mr. Sharpless designed some Eniac equipment."

73. Bruno Latour and Steve Woolgar, in *Laboratory Life* (Beverley Hills, Calif., 1979), 219, point out that "a key feature of the hierarchy is the extent to which some people are regarded as replaceable."

74. Rupp (n. 4 above), 161.

75. Ibid., 161–62.

76. U.S. Department of Labor, "The Outlook for Women in Mathematics and Statistics" (n. 8 above), 9–11. See also U.S. Department of Labor, "A Preview as to Women Workers in Transition from War to Peace," *Women's Bureau Special Bulletin*, 1944; Rossiter, *Women Scientists in America: Before Affirmative Action, 1940–1972* (n. 9 above), chap. 2.

77. U.S. Department of Labor, "The Outlook for Women," 11.

78. Army Service Forces Office of the Chief of Ordnance, Washington, D.C., to personnel at BRL, 29 January 1946, Goldstine Papers.

79. Herzenberg and Howes (n. 10 above).

80. U.S. Department of Labor, "Employment Opportunities for Women Mathematicians and Statisticians," *Women's Bureau Bulletin* 262 (1956): vi.

81. For these women's later employment histories, see Fritz (n. 2 above), 17.

82. Margaret Rossiter, "The Matilda Effect in Science," *Social Studies of Science* 23 (1993): 325–41.

83. U.S. War Department, *You're Going to Hire Women*, booklet produced to persuade managers and supervisors to hire women, cited in Chester Gregory, *Women in Defense Work During World War II: An Analysis of the Labor Problem and Women's Rights* (New York, 1974), 12.

84. For example, the *Women's Bureau Bulletin* 262 (1956) features several pictures of women working with computers and mentions women coding and programming.

85. Gerda Lerner, "The Necessity of History," in *Why History Matters: Life and Thought* (New York, 1997), 119.

Spotlight on Emma Barth, P.E.

A Typical Woman Engineer?

Lauren Kata

EMMA C. BARTH [*1912–1995. ED.*] WAS PART OF A GEN-
ERAtion of women who faced unprecedented employment
opportunities as a result of men's recruitment into combat
positions during World War II. Wartime circumstances created new organizations and programs specifically designed
to recruit women into service. In the call for womanpower,
science and engineering played a central role.

It was in this atmosphere that Emma Barth capital-
ized on her interest in engineering. Prior to the war, during
the Depression years, she received her education from the
University of Pittsburgh, earning both a B.A. (1931) and
an M.A. (1937) in German, and an M.A. in education.
Employment at that time, however, was scarce: the only
teaching work available was part-time work in evening and
summer schools. Anxious for full-time work, and interested
in engineering because that was her brother's field of study,
Barth enrolled in drafting courses at the Pittsburgh Aeronautics Institute in 1942. She secured a job as a draftsman
for the H.J. Heinz Co. that same year.

Kata, Lauren. (2002). "Spotlight on Emma Barth, P.E.: A typical woman
engineer?", *SWE Magazine*, February/March, 29–31. Reprinted with
permission of the Society of Women Engineers.

Emma Barth.
SWE Archives, Walter P. Reuther Library, Wayne State University, used with permission.

For both women educated in engineering before the war, and women like Barth who took courses as part of a wartime training program, wartime labor shortages created opportunities to apply technical skill in the workplace. As a draftsman for Heinz, Barth worked in the canning plant, which was temporarily converted into a drafting office, and there she drafted wooden wings for gliders. Following her position at Heinz, she joined Westinghouse Electric as a draftsman in 1944, in the Turbine Generator Division at Westinghouse's East Pittsburgh Office.

For many women, as the general story goes, the new wartime positions did not extend up the job ladder, especially when male veterans returned from the war. If women were retained in technical jobs, they tended to remain at the assistant level. This was true even for women who had received their education in engineering, science, or mathematics before the war.

Emma Barth's experience as a draftsman for Westinghouse, working in industry, inspired her to pursue a degree in engineering, in the hopes that she could remain employed in the field. When she began her job in 1944 there were 18 women in her department—which she believed was a fairly common number of women in drafting at the time. Barth, however, had hopes for advancing out of the women's drafting department. On the advice of a friend in the engineering department, she enrolled in engineering classes at the University of Pittsburgh in 1944. Paying for the classes herself, she went to school in the evening for seven years and in 1951 graduated with a B.S. in general engineering—the only female in her class and the first woman to graduate in engineering from the University of Pittsburgh's evening school. *[The following excerpts from Barth's diary illustrate her long days in pursuit of an engineering degree. Ed.]*

Monday, October 1, 1951

I got up feeling fine, hardly napped on the train at all and got to work pretty promptly. But at work I started to feel sleepy and weary and shaky, and never did snap out of it.

The rooms looked torn up and rearranged. Brenner is now against the window behind us, with Carl J. next to him. Ward didn't show up— "sick" was the word we got. P. remarked, "The 49-hr weekend must have been too much for him." So I signed off some of his tracings, and then made a determined attempt to get a S.O. out of the way. It shouldn't have been too difficult since it is to be a duplicate, but now I smack into the problem of having ordered incorrect brush riggings. . . . That means a detour to check status, cancel, schedule new, etc.

By 5:00 was too dizzy-sleepy to do anything other than sleep on the train—ate in the Pitt cafeteria which they have this year reorganized to get away from the one-price meal. There are now single items which, of course, add up—I found myself with a bill of $1.02. Of course, the tomato juice must have boosted the total considerable. And didn't even taste particularly good, either.

Met Mrs. T. of the Penn Mutual Co., by appointment at 7:00. I had to tell her that Dad . . . wanted me to put my extra money into a B&L instead. She agreed that if I could do better that way, it would be wiser!

Then, leisurely made my way up to Thaw Hall—and again found myself so tired that I wanted to collapse into deep sleep. It was good that I can get to Room 103 early enough to have some kind of seat. There seem to be over 30 in the Lubrication class—all of the fellows strangers.

Dr. Boyd talked interestingly and clearly. But towards the end I was getting too sleepy to pay close attention. At the end of class I went up and gave him the "vital statistics" of degree, occupation, etc. He still has no class cards. He repeated his statement of "This is the first time this has happened."

Prompt st. car service got me home at 10:30. But till I glanced at the paper, tried a bit of voice practice, and did a rare minimum of urgent things, it was midnite.

As soon as she received her degree, Barth was promoted from draftsman to associate engineer in the mechanical design section where she continued to work on turbine generators. In 1960 she was promoted to engineer in Westinghouse's large rotating apparatus division. She was later promoted to advanced engineer in 1975. Barth was employed at Westinghouse from 1944 until she retired in 1977.

An ambitious engineer, Emma Barth was committed to professional development. She was a founding member and later president of the SWE [Society of Women Engineers. Ed.] Pittsburgh Section, and was also the first editor of the SWE Newsletter. In 1952, she became an engineer-in-training and joined the Society of Professional Engineers. She obtained her Pennsylvania Professional Engineers License in 1961, and after she retired in 1977, she became a life member. For her years of service and encouragement to young women, she received the Westinghouse Community Service Award in 1977, the first woman to receive that award. Engineering wasn't her only passion however; she was a serious drama student, who practiced voice exercises daily and participated in a local theater group. [Further excerpts from Barth's diary describe some of her interests outside of her job. Ed.]

Tuesday, October 16, 1951

Managed to get up this morning, wasn't too much of a struggle.

Wore the gray suit to work for the first time. . . .

. . . It seems to me I spent considerable time talking over SWE today, but with happy results. Sandy called to tell me that three women from Chicago were coming in for Saturday! She is all thrilled now at the prospect of our having members in Chicago who will be on the ground floor for handling details of the Chicago convention!

In the afternoon I called Mgt [Margaret] Kearney and discovered that she is still very much interested in joining. She had a check and a letter all ready to send, but she has a substitute typist this week, and the girl had made some typing errors! However, Kearney is still on our side, and is going to work hard to increase our membership. Much of the rest of the day I spent in checking thru the hodgepodge of changes in the TVA job. . . .

. . . Got to Mrs. Lissfelt's in time—she did admit that the placement of my voice has improved a lot since last time. But the way I had been singing the E still wasn't completely right. Still, she was pleased with the way I pick things up. Now if only I had time for a bit of rehearsal during the weeks so that I could take full advantage of my 5-dollar lesson. . . .

. . . Just taking care of the minimum essentials before going to bed lasted until 1:30!

Thursday, July 3, 1975

. . . I should have been almost deliriously happy at the prospect of the four unbroken days, but I felt as draggy as usual as I walked down from the bus stop. It was looking very gray and gloomy and for a good reason—we soon had a crashing storm, which brot [sic] rain in on my windowsill and set the old administration bldg. on Duquesne University campus on fire with a stroke of lightning.

After supper I decided skimming thru magazines could be combined with watching some TV programs. "Growing up Female" on WQED sounded most intriguing. It was the showing of a film made 1970, being viewed by an assorted group of women in one room, and by a group of men in another. Then we had the excited reactions of the women to the film, followed by the sometimes horrifying reactions of the men, many of whom couldn't understand "what do women want to" Also, the women were secretly watching the men, and we were then allowed to view their reactions. Finally, all 16 or so men and women were put together, most of them jabbering excitedly. (I don't know why they don't have a moderator for some such confrontation.)

Anyway, they warned you "Be watching this with somebody else because you'll want to discuss it afterwards." I couldn't, of course. And then Bertha Bailey called in the midst of it, to thank me for her birthday card!

I finally took myself downstairs to type the letter to Mrs. Post at Old Economy. By 11:00 it wasn't done yet, and I was getting sleepy. So then I went up and watched the news from the couch. I'm afraid I napped a bit toward the end.

Finally got the letter typed and then got to bed, sans curlers and definitely without setting an alarm.

Wednesday, July 30, 1975

Before I finished breakfast, there was a phone call from Ralph, asking for a good number!
Back to the bus routine again. Napped a bit, and felt positively chipper as I started looking into a new I.B. Had to get some more info from Murphy. Then Cunningham told me the details of his woes in trying to get an operator's manual thrown together on short notice. Trafford didn't have any prints. But I had ordered 75 each of these particular drawings four years ago!
Called Robshaw and finally got to him before lunch. He said that the prints were all filed in the general file by dwg number, not kept separate. Oh-my-gawd!! When I told Urbansky to scrap outline supplements on his list of dwgs. I was having the operators manual illustrations thrown out again!! Maybe there is still hope, since I have gone thru only about a fourth of the dwgs sent to me as samples. . . .

Barth's diaries offer a well-rounded perspective of a woman engineer's daily life—which included technical concerns, networking, social events, artistic passion, and ordinary human activities. Her life as a woman engineer is humanized by her own words; rather than being a statistic, or being categorized as another woman first, in her own words the diaries allow her to be Emma Barth. As historians and sociologists and engineers continue thinking and writing about women engineers, cultural and humanistic elements are essential for a true understanding of the individual's role in the history of a collective group.

Some of Barth's reflections and experiences may sound familiar, making them representative, ahistorical and even timeless. The diary entries, however, are also the product of the time in which they were written. With these thoughts, was Emma Barth a typical woman engineer? Is there such a thing as a "typical" woman engineer—and how is that measured? Historically, as changes within engineering, the economy and society have taken place, how have these changes shaped both the image and the experiences of women engineers? It is particularly fitting to ask these questions during Women's History Month. The Emma Barth Diaries will remain a powerful primary resource for further analysis and a key to answering some of these questions.

NOTE: All transcriptions of diary entries by Lauren Kata. Information used in this article, including the Emma Barth Diaries, came from the Society of Women Engineers Archives. For more information on the SWE collections, see "Making SWE History," SWE Magazine, (December 2001/

January 2002). Or, contact the SWE archivist at the Reuther Library: (313) 577-9373.

This article was also informed by the work of others. For further reading:

Juliet K. Coyle, "Evolution of a Species—Engineering Aide," in U.S. *Woman Engineer*, (April 1984), pp. 23–24.

Jennifer S. Light, "When Computers Were Women," in *Technology and Culture*, 40.3 (1999), pp. 455–483.

Ruth Milkman, *Gender at Work: The Dynamics of Job Segregation by Sex During World War II.* (Chicago, 1987).

Ruth Oldenziel, *Making Technology Masculine: Men, Women, and Modern Machines in America, 1870–1945.* (Amsterdam, 1999).

Ruthann L. Omer, EIT. "Emma Barth, PE: One of the First," in PE *Reporter*, Feb/March 1985.

Margaret Rossiter, *Women Scientists in America, Volumes 1 and 2* (Baltimore, 1982; 1995).

Leila Rupp, *Mobilizing Women for War.* (Princeton, 1978).

13

1998   *Flight Path*

Betsy Flagler

A ROAD WINDING PAST NC [NORTH CAROLINA] STATE'S
College of Engineering is named in honor of an aerospace
pioneer who has paved the way for so many other women—
Katharine Stinson [1917–2001. Ed.], the first woman to
graduate from NC State's engineering program, received a
degree in 1941, then began an illustrious 32-year career in
aircraft safety.

When NC State renamed a North Campus street Stinson Drive in 1997, it represented quite a turnaround from
the school's initial reaction to Stinson six decades ago. NC
State said no to Stinson's attempt to register as a freshman,
but eventually she broke through that barrier and has been
an inspiration ever since.

"Over the years, she has been very generous about
coming back to talk to engineering students," says long-
time friend Frances D. "Billie" Richardson, the first woman
on NC State's engineering faculty. "She is always such an
inspiration."

After finally being admitted into NC State as a junior,
Stinson had no time to worry about being the only female
learning how to dismantle airplane engines or use a slide
rule. In her efficient manner, the Raleigh native was too
busy attaining a goal inspired by her own heroine, aviatrix
Amelia Earhart.

"I was 15, working with airplanes and learning to
fly, when I met my idol at the Raleigh airport," Stinson

Flagler, Betsy. (1998). "Flight path." *NC State Magazine*, Spring, 10–15.
Reprinted with permission of the author and *NC State Magazine*.

Katharine Stinson.
SWE Archives, Walter P. Reuther Library, Wayne State University, used with permission.

recalls during an interview in her home in Pinehurst. Earhart was flying to promote Beechnut gum, and her plane was being serviced when Stinson got the chance to meet her.

"I told Miss Earhart I wanted to be a pilot, but she said just being a pilot wouldn't be enough to make a living. She said I should study aeronautical engineering."

The teenager wasn't sure what Earhart meant, but she took off on a mission to find out—and never looked back.

"She has become a symbol women can succeed in engineering" says another longtime friend, Mary Yionoulis, former director of engineering communications at State. "Her love for her career, as well as for education, has inspired many women students throughout the years."

Yionoulis has kept abreast of Stinson's accomplishments since the two met in 1971, when Stinson became the first woman to receive the NC State Alumni Association Distinguished Engineering Alumnus award and the first woman to serve on the NC State Alumni Association Board of Directors.

The citation for her Distinguished Engineering Alumnus award says: "Her versatile background as a well-qualified pilot, her knowledge of engine and airframe maintenance, plus her proficiency as an aeronautical engineer have made this outstanding woman engineer an invaluable asset to this nation."

Stinson always knew that if she were going to be an aeronautical engineer, NC State was where she needed to get an education. But when she tried to register as a freshman, she was told that she belonged at Women's College in Greensboro, now the University of North Carolina at Greensboro.

"Little girl, what are you doing here?" Stinson recalls Dean Wallace C. Riddick saying, "I have to be an engineer because Amelia Earhart told me to be," replied Stinson.

Riddick told her she could enroll when she became a junior. She pushed him to tell her how many credits she would need, then headed to Meredith College to regain a scholarship that she had turned down. She crammed the required 48 credits into one year—with two-thirds of a credit to spare. Then she was back in the registration line at State, ready to enter "the man's world" of engineering.

"It bothered the professors that I was there, but the boys were great," she says. "Some of the faculty had the idea that girls didn't do things like engineering, and, of course, they didn't at that point. But I saw no reason they couldn't."

Stinson proved to be among the top students and received an award from the Order of Saint Patrick, a national engineering fraternity that honors outstanding seniors. She chuckles about an error on the award certificate, which was designed for men only—as was the rest of the school:

"Katharine Stinson, having proved himself an engineer and loyal follower of Saint Patrick, is hereby dubbed: Knight of Saint Patrick."

After graduation, Stinson was hired by the Civil Aeronautics Administration as its first female engineer. The agency evolved into today's Federal Aviation Administration.

"I never thought about the fact that I was the only woman, because I never saw any women," she says. "That may sound funny, but I just wanted to be a good engineer."

During her career, "Call Katy" became a frequent solution when complex engineering problems arose. As a well-respected safety warden of the skies, she developed procedures to ensure that proposed aircraft designs either met or surpassed government safety standards.

After graduation, Stinson was thrown into challenges quickly because of the shortage of engineers during World War II. One of her first projects was to convert light airplanes into gliders for pilot training during the war.

She also gets credit for revising procedures to handle airworthiness directives about unsafe conditions on aircraft.

"For many years I wrote every one of them," recalls the soft-spoken Stinson. "Whenever we found anything wrong with an airplane, we would notify the owners by telegram. We prided ourselves on the prevention of accidents."

Stinson retired in 1973 but has remained active in many organizations, including The Ninety-Nines, an international group of licensed female pilots that Earhart founded in 1929. As a member of the Society of Women Engineers, which she helped to organize and lead, and the Soroptimist International of the Americas, a worldwide service organization for professional women, Stinson has encouraged women to attain their goals. She gave the keynote address at the Soroptimists' 75th anniversary in December.

"Being a woman should not deter you from being an engineer," she says. "Don't think that it's glamorous. It's hard work. The main thing is to really want to do it and work hard. If women really want to, they can do anything they want. I see unlimited opportunities."

As a child, Stinson built model airplanes out of balsa wood and rubber bands, or fished in a creek on the Fuquay-Varina farm where she grew up. She loved to listen for the occasional plane flying overhead.

"I was a little country girl catching little country fish and dreaming about airplanes," Stinson says.

Stinson—the only one in her family who was interested in aviation—scoured the newspaper for photos of planes and stories about her heroes. Her interest paid off when she got the chance to fly for the first time at age 10. One of her heroes, aviation pioneer Eddie Stinson (no relation), took her up in his enclosed cockpit plane while he was on a business trip to Raleigh in 1927.

"He didn't ask my mother," Stinson recalls. "He just picked me up and put me in the plane. That flight was a dream come true—that and meeting Amelia Earhart."

Stinson's first flight hooked her on the idea of becoming a pilot. Five years later, Earhart inspired her to set an even higher goal: a technical degree combined with a pilot's license.

In 1937, as a gutsy college student pursuing that goal, Stinson was able to talk to her idol a second time one afternoon in Washington not long before Earhart's final journey. When the aviatrix was declared missing, Stinson stayed awake listening to news reports on the radio at Meredith.

"She was my heroine," Stinson says. "Nowadays, I guess they call them mentors."

Tiny colored pins on a wall map in Stinson's den mark the nearly 200 countries she has visited. She's had a slight stroke that has slowed her traveling, but she still hopes to get to Antarctica, the only continent she has yet to visit.

Stinson is an avid golfer. She can walk out the back door of her Pinehurst home and onto the course, starting at the sixth fairway. "I call that my fairway," she says as she looks from her enclosed sun porch.

Stinson, who maintained a pilot's license for 50 years, selected the retirement community in part because golfers are not required to use golf carts but instead may walk the course.

"As I got closer to retirement, I'd think about where I wanted to live. I looked all over the world as I traveled, and I came back here," she says. "I can go right to the tees and play my own round and end up back home. All the years I worked, I didn't have time to play golf. When I was at State, I didn't have the money."

She has so many friends who visit that she owns the home next door and uses it as a guest house. For her 80th birthday, friends from as far away as London joined her for a celebration in her home. She lives only about an hour from Raleigh, which makes it easy for her to return to see friends, speak to groups or attend class reunions.

"Katharine is keenly interested in the College of Engineering, particularly in seeing that women have an opportunity to excel," says Ben Hughes, executive director of development and college relations at the College of Engineering. "She has led by example."

The current dean of the College of Engineering, Nino A. Masnari, says Stinson has brought tremendous pride to the institution that initially tried to refuse her admission.

NC State's director of gift planning, Joan DeBruin, has been so inspired by Stinson that she took her daughter to Pinehurst one afternoon to chat with the aerospace pioneer.

"I thought it would be good for my daughter, who is also in engineering, to hear Katharine talk about the challenges she has faced," DeBruin says. "Her dedication to the profession of engineering rubs off on you. She has been an inspiration to anyone who has ever met her."

In 1987, Stinson established a scholarship to help women engineering students achieve their goals. Recipients are called "Katharine Stinson Scholars."

"Katharine has a huge love for NC State," DeBruin says. "She has wanted to give back to future students so they don't have to go through what she went through."

Stinson also has set up a charitable remainder trust to benefit her existing scholarship and to start a new one in honor of Coach Kay Yow. As the first full-time women's basketball coach at NC State, Yow says she identifies with Stinson's efforts.

"Katharine is a source of wisdom and strength because she has been through the pioneering efforts and come out a winner," says Yow, who invited Stinson to be the honorary coach at her team's season opener. "I want my players to know the part she played at NC State. She paved the way. My players are winners just knowing her."

After Katharine Stinson managed to get in the door at NC State, a *Raleigh News and Observer* reporter interviewed the 20-year-old about her interest in flying and aeronautical engineering.

"Why do people want to know about me?" Stinson asked the reporter, "I haven't done anything yet—maybe I never will. But I really ought to do something first before they are interested in me."

Her many achievements and contributions since then include these highlights:

- In 1941, Stinson became the first woman to receive an engineering degree from NC State, where she majored in mechanical engineering with an option in aeronautics. She was one of only five women in the nation to graduate that year with an engineering degree.
- The Civil Aeronautics Administration hired her in 1941 as its first female engineer. She worked for the agency, now the Federal Aviation Administration, for her entire 32-year career.
- One of her first assignments as a CAA engineer was to convert light airplanes to gliders for pilot training during World War II. She also figured out how to reconvert the trainers back to engined airplanes after the war.
- From 1964 until her 1973 retirement, she was the technical assistant chief in the FAA's Engineering and Manufacturing Division, which establishes design and safety standards and approves aircraft used in civil operation in the United States.

Other Accomplishments:

- 1953–55: Helped organize and was third president of the Society of Women Engineers.

- 1961: Received FAA Sustained Superior Performance Award.
- 1966: Tapped as an honorary member of Phi Kappa Phi at NC State.
- 1970: Installed as president of the Soroptimist International of the Americas, a worldwide service organization for executive and professional women.
- 1971: Received NC State Alumni Association Distinguished Engineering Alumnus citation, becoming the first woman to receive that honor.
- 1971–74: Became first woman to serve on the board of directors of the NC State Alumni Association.
- 1984: Received Distinguished Women in the Aerospace Industry Award.
- 1987: Named Aviation Pioneer of the Year by the Institute of Aeronautics and Astronautics.
- 1987: Set up the Katharine Stinson Scholarships for Women in Engineering at NC State.
- 1997: Through a charitable remainder trust, established a $100,000 scholarship in honor of NC State women's basketball coach Kay Yow.
- 1997: Stinson Drive named for her on NC State's North Campus,
- 1997: William Friday featured his former State classmate on a segment of "North Carolina People" and introduced her as a "bona fide pioneer."

14

2003 · *Women Leaders in Engineering Societies*

Peggy Layne, P.E.

THIS IS A BANNER YEAR FOR WOMEN LEADERS IN THE largest engineering societies in the United States. LeEarl A. Bryant, P.E., just completed her presidency of the Institute of Electrical and Electronic Engineers-USA, the geographic unit that represents almost ⅔ of IEEE's 377,000 members worldwide. Susan H. Skemp is currently president of the American Society of Mechanical Engineers (ASME), and Diane Dorland, P.E., is president of the American Institute of Chemical Engineers (AIChE). Waiting in the wings are Patricia D. Galloway, P.E., of the American Society of Civil Engineers (ASCE) and Teresa A. Helmlinger, P.E., of the National Society of Professional Engineers (NSPE), who will begin their terms as president later this year. Skemp is the second woman to lead ASME, while the other four are (or will be) the first women presidents of their respective organizations.

ASCE is the oldest of American engineering societies, founded in 1852. ASME followed not long after in 1880. IEEE is a successor organization of the American Institute of Electrical Engineers, founded in 1884. The chemical engineers organized in 1908, and NSPE is a relative latecomer, founded in 1934. Women have worked as and received

Layne, Peggy. (2003). "Women leaders in engineering societies." *SWE Magazine*, Spring, 20–22, 24. Reprinted with permission of the Society of Women Engineers.

Women engineering society presidents, clockwise from left: Diane Dorland, Susan Skemp, LeEarl Bryant, Patricia Galloway, Theresa Helmlinger.

degrees in engineering since the late 1800s, but were not welcomed as members of most engineering societies until well into the twentieth century. Although we still have a long way to go, the tremendous progress that women engineers have made in the last 25 years is exemplified by these five women and their current positions as highly visible leaders of the profession.

All of the five women featured in this article are current members of the Society of Women Engineers (SWE). Three have served as section presidents—Bryant, Galloway, and Helmlinger—two are recipients of SWE's Distinguished New Engineer award—Galloway and Helmlinger—and one, Bryant, is a SWE Fellow. Dorland supports the SWE student section at Rowan University, where she is dean of engineering. Skemp joined SWE recently, at the behest of then-president Roberta Gleiter. Galloway notes "SWE helped me learn what being a leader in a professional organization is all about and served as an important foundation to my ASCE leadership roles. SWE assisted me in building professional confidence as well as serving as a resource and network of other women engineers that were few and far between in ASCE, especially during my early years of involvement." Helmlinger agrees, stating "SWE has inspired my commitment to the profession, plain and simple."

LeEarl Bryant has been a member of IEEE since her student days at Texas Tech, where one of her professors would have pop quizzes following IEEE student branch meetings. "Unlike most other students," Bryant says, "I knew of the organization because my father was a member." While IEEE's 377,000 members reside in 150 countries, the IEEE-USA was established in 1973 to promote the careers and public-policy interests of the organization's more than 230,000 U.S. members.

Bryant began her career at Collins Radio Company in Dallas, and spent much of her career as a technical manager with an extensive background in telecommunications, transportation, and defense industries. Bryant also holds a master's degree in electrical engineering with a biomedical option from Southern Methodist University. While serving as an IEEE Congressional Fellow in 1993, Bryant worked for Fort Worth's Representative Pete Geren on telecommunications, technology, competitiveness, health, education, and aging issues.

As president of IEEE-USA, Bryant focused on government policy issues that affect engineers as well as issues such as pre-college initiatives and professional development. One of Bryant's priorities during her presidential year was to improve the image of engineers. She launched a new project to develop 30-second television spots highlighting the contributions of engineers to society.

On the policy side, Bryant points out "an issue of great concern to me personally addresses the present unemployment status of so many American engineers at a time when our government supports and encourages continual entry of engineers outside of the U.S. as foreign temporary workers." Bryant represented IEEE-USA at several meetings that discussed issues related to the U.S. engineering workforce, from encouraging more young people to pursue engineering careers to ensuring that engineers maintain their professional skills and have opportunities to grow throughout their professional lives. "We need to ensure that pre-college students are better prepared to enter the workforce and contribute to society," says Bryant.

Diane Dorland was introduced as president of AIChE's 50,000 plus members at their annual meeting last November. Dorland's career has spanned industry and academia, with time out to raise a family. She joined AIChE as an undergraduate at the South Dakota School of Mines. "AIChE provided the networking backbone and technology support that I needed throughout my career." Dorland worked for Union Carbide and DuPont before taking time out for motherhood and earning a Ph.D. at West Virginia University. She then moved into academia, joining the faculty at the University of Minnesota, Duluth, where she became chair of the chemical engineering department. From Duluth she joined Rowan University, in Glassboro, New Jersey, as dean of engineering in 2000. In addition to AIChE, Dorland has served in leadership positions of NSPE, the American Society for Engineering Education, and the Water Environment Federation.

In setting priorities for her year as president, Dorland is focusing on the expansion of chemical engineering beyond the traditional petrochemical arena. "Chemical engineering influences the fabric of the workplace and our society in many ways. Chemical engineers are building on their educational foundations but branching in career areas such as business, medicine, law, electronics, finance and biotechnology, to name just a few. AIChE is positioned to expand its influence beyond the traditional chemical process industries and capture the energy of this changing nature of the profession. It is my goal to guide AIChE towards increased professional awareness of this broader influence, truly making it our lifetime professional home."

Sue Skemp, 2002–2003 president of ASME International, has been involved in that 125,000-member organization for 21 years, working in all five of ASME's councils. Skemp earned her Bachelor of Science degree as a reentry student at Florida Atlantic University. "I first joined ASME as a student in 1978, when I returned to college as a non-traditional student after my children were born." She was asked by a professor to participate in hosting an ASME regional student conference, and "the hook was planted." On graduation, Skemp joined Pratt & Whitney and became active in the Palm Beach Section of ASME. She has spent her entire career at Pratt & Whitney, where she is currently manager of advanced technology planning, and has worked on the development of aircraft and rocket engines for the military, space, and commercial markets.

Halfway through her term as president, Skemp has emphasized a number of areas, all with the intent of improving value for ASME's customers, members, and the engineering profession, addressing the education pipeline, and ensuring the viability and growth of ASME in the future. Her goals include public policy, globalization and inclusiveness, ethics and professionalism, and pursuit of emerging technologies.

One of the activities Skemp prioritized for her year as ASME president was a series of programs with the Girl Scouts as part of "Introduce a Girl to Engineering" during National Engineers Week, February 16–22, 2003. Skemp worked with Girl Scouts in Connecticut, Ohio, and Washington D.C. to help them earn credit toward their "Building a Better Future" patch. Activities included the 10-minute video, "Mothers of Invention," as well as hands on activities. "Sharing our personal experiences as women in engineering with our target group of Cadette and Senior Girl Scouts in grades six through 12 will help them realize the myriad of opportunities available to them in our field," Skemp says, urging other women members of ASME to join her in outreach programs.

Terri Helmlinger will take office as president of the 60,000-member NSPE in July 2003. She earned her B.S. in engineering operations from North Carolina State University, where she currently serves as assistant vice-chancellor for extension and engagement and executive director of the industrial extension service. Helmlinger also holds an M.B.A. from Duke University. She began her career at Carolina Power and Light, now Progress Energy, where she held many positions over almost 20 years, including director of commercial and industrial market development.

Helmlinger became active in NSPE at the urging of her first boss, who encouraged her to become a registered professional engineer and to get involved in professional society activities. After serving in leadership roles at the local and state levels of the Professional Engineers of North Carolina, Helmlinger moved up to the national level as a director and regional vice-president before becoming president-elect last year. In addition to her NSPE activities, Helmlinger served as president of the North Carolina Section of SWE (now the Eastern North Carolina Section). On a personal note, she first recruited the author to become involved in SWE as an officer in that section.

Change is one of the themes for Helmlinger's year as president of NSPE. "NPSE is transforming—so much so that it requires a chain of leadership to bring us completely through the transition. I want my year as president to be a strong link in that chain. My personal strengths that I can contribute will be focused on our image campaign and on our efforts to collaborate with other engineering societies for some of the overarching efforts our profession requires."

Pat Galloway becomes president of ASCE in October 2003, after serving as director, district representative, and chair and member of numerous committees in that 125,000-member organization. A Purdue alumna, she earned her bachelor's degree in three years, and received Purdue's Distinguished Engineering Alumni Award in 1992. She also holds an M.B.A. from New York Institute of Technology. Galloway began her career with CH2M Hill in Milwaukee and subsequently joined the Nielson-Wurster Group, an international engineering and management consulting firm where she is currently president and chief executive officer. She is an expert in construction scheduling, project controls, and dispute resolution.

After joining ASCE as a student at Purdue, Galloway served as president of the concrete canoe club. One of her first supervisors at CH2M Hill encouraged her to get involved in ASCE at the professional level, and helped her obtain an appointment to a national committee on construction management. She set her sights high, working toward the presidency of ASCE since joining the organization as a student. In addition to her involvement in ASCE, Galloway has served as president of the Wisconsin and New York sections of SWE.

Galloway's goals for her year as ASCE president address globalization, public policy, ethics and professionalism. "I also want to serve as a role model to both young engineers and women to demonstrate that it is not just an 'old boys' network and if you work hard and want to contribute to your profession that you can excel to the top" she says. "Being in the construction industry, I also hope to bring back the construction industry engineers and professionals that have left ASCE since there has been a long-time perception that ASCE is a 'design' oriented professional organization."

In terms of leadership style, these women see themselves as consensus builders, but also note that different styles of leadership can be appropriate in different situations. Continually learning new skills themselves, these leaders also strive to encourage learning and skill development in those around them.

Bryant focuses on obtaining "buy-in" from others. "I try to lead so that others get to influence and hopefully have buy-in to the idea or project and method of approach. In some cases, I'm probably not as democratic as I normally would want because my mind may be committed to a project or idea. I do try to be flexible so that others can see that they've influenced the approach, outcome, or project in some way."

Skemp emphasizes making decisions based on the best knowledge available. "I challenge the status quo, and work to inspire others to question 'how can we do it better,' pushing the frontier. I believe in establishing a vision that can anticipate near term achievements, balanced with long term possibilities, somewhat riskier but creating opportunities for the future. Teams work best when the members feel that they all have an opportunity to contribute, rather than be dictated to. I also believe that identification of future talent is extremely important in ensuring the viability of professional organizations such as ASME. To that end, I have made a point to tap into not only current expertise, but also the upcoming leaders to ensure that inclusiveness is not just a buzz-word for ASME."

Galloway stresses the importance of attitude and modeling the behaviors expected of others. "With respect to the managers and other senior officers in my company, I consider myself a person of true democracy and always attempt to arrive at consensus after discussion as I firmly believe it is important for all the managers to buy-in to the concepts presented. Regarding the staff, I feel it is very important to lead by example. I would not expect others to work extensive overtime or to reflect a positive attitude unless I did so myself. I am a 'do as I do' and not just 'do as I say' person. I have found this to be a team-building concept that has worked very well. I also keep a positive attitude even in the worst of times as a positive attitude in my experience always wins out and results in a positive response and result."

Dorland observes, "I have seen that leadership is frequently situational. What has worked in one instance doesn't necessarily work at another time or place. I am continually developing new leadership skills and new understanding of the leadership process. That said, I tend to be a consensus builder. My goal in any position is to leave a legacy of successful operation and an empowered cohort that will manage transition and new development well."

According to Helmlinger, "I used to consider myself solely a democratic or participative leader, but I have come to realize in the last two years that I can often be a situational leader whereby I can change my style to fit the situation. In my most recent roles, both at work and in my volunteer leadership, I am fortunate to have the opportunity to roll these styles up into what I have come to term a 'meta' leader—a leader of leaders."

SWE Magazine asked these outstanding women for any "words of wisdom" for our readers:

Bryant observes, "When life isn't going the way you dream or desire, remember that this too will eventually pass. When you encounter the first career or life setback due to increasing age, remember that everyone gets older, and so will those who cause your problems. When you wonder if a career in engineering is right for you, ask yourself whether you like the challenges, using your brain, and the idea of creating the future. If so, the answer is probably yes. When you think that other careers may have fewer ups and downs, it's not necessarily so. When you think that other fields of study may require less work, you're probably right!"

Skemp uses the analogy of a tapestry for her career. "No one individual or organization can accomplish everything alone. Like a tapestry, the threads of value and knowledge base are interwoven. Alliances and coalitions are needed to provide the greatest value in reaching the pipeline, challenging the technology frontier, and meeting the requirements and expectations of our constituents—members, employers, and the public at large." She also reminds readers to "have fun!"

Galloway notes "If someone tells you that being president did not take a lot of hard work and personal sacrifice, then I would be willing to bet that that person is not only NOT a president, but will never be one." Her keys to success are "Communication, Confidence, and Commitment. If you have these 3 C's, there is nothing that you cannot do if you really want to and put your mind to it."

Dorland's advice is "to face the challenges with a smile and a sense of humor, think beyond the context of the immediate situation, and respect yourself."

Helmlinger provides some words of encouragement. "Women engineers should be proud of the contributions we all make every day—we are enriching our profession in more ways than any one article can illustrate."

Further Reading

PROFILES OF WOMEN ENGINEERS:

Hatch, Sybil. (2006). *Changing our world: True stories of women engineers,* American Society of Civil Engineers Press, Reston, Va.

ANTHOLOGIES ABOUT WOMEN IN SCIENCE,
SOMETIMES INCLUDING ENGINEERING:

Abir-Am, Pnina G., and Outram, Dorinda, eds. (1987). *Uneasy careers and intimate lives: Women in science, 1789–1979,* Rutgers University Press, New Brunswick, N.J.

Ambrose, Susan A., Dunkle, Kristin L., Lazarus, Barbara B., Nair, Indira, and Harkus, Deborah A., eds. (1997). *Journeys of women in science and engineering: No universal constants,* Temple University Press, Philadelphia, Pa.

Haas, Violet B., and Perrucci, Carolyn C., eds. (1984). *Women in scientific and engineering professions,* University of Michigan Press, Ann Arbor.

Kass-Simon, G., and Farnes, Patricia, eds. (1990). *Women of science: Righting the record,* Indiana University Press, Bloomington.

BOOKS ABOUT WOMEN IN SCIENCE,
SOMETIMES INCLUDING ENGINEERING:

Eisenhart, Margaret A., and Finkel, Elizabeth. (1998). *Women's science: Learning and succeeding from the margins,* University of Chicago Press, Chicago.

Etzkowitz, Henry, Kemelgor, Carol, and Uzzi, Brian. (2000). *Athena unbound: The advancement of women in science and technology*, Cambridge University Press, Cambridge.

Gornick, Vivian. (1983). *Women in science: Portraits from a world in transition*, Simon and Schuster, New York.

Preston, Anne E. (2004). *Leaving science: Occupational exit from scientific careers*, Russell Sage Foundation, New York.

Rossiter, Margaret. (1982). *Women scientists in America: Struggles and strategies to 1940*, Johns Hopkins University Press, Baltimore, Md.

Rossiter, Margaret. (1995). *Women scientists in America: Before affirmative action 1940–1972*, Johns Hopkins University Press, Baltimore, Md.

Wasserman, Elga. (2000). *The door in the dream: Conversations with eminent women in science*, Joseph Henry Press, Washington, D.C.

BOOKS ABOUT WOMEN AND GENDER IN ENGINEERING AND TECHNOLOGY:

Canel, Annie, Oldenziel, Ruth, and Zachmann, Karin, eds. (2002). *Crossing boundaries, building bridges: Comparing the history of women engineers 1870s–1990s*, Harwood Academic Publishers, Amsterdam.

Carter, Ruth, and Kirkup, Gill. (1990). *Women in engineering: A good place to be?*, MacMillan, London.

Hacker, Sally. (1989). *Pleasure, power & technology: Some tales of gender, engineering, and the cooperative workplace*, Unwin Hymen, Boston.

Hacker, Sally L. (1990). *"Doing it the hard way": Investigations of gender and technology*, Dorothy E. Smith and Susan M. Turner, eds., Unwin Hyman, Boston.

McIlwee, Judith, and Robinson, Gregg. (1992). *Women in engineering: Gender, power, and workplace culture*, State University of New York Press, Albany, N.Y.

Oldenziel, Ruth. (1999). *Making technology masculine: Men, women, and modern machines in America 1870–1945*, Amsterdam University Press, Amsterdam.

Reynolds, Betty, and Tietjen, Jill. (2001). *Setting the record straight: The history and evolution of women's professional achievement in engineering*, White Apple Press, Denver, Colo.

ANTHOLOGIES OF FEMINIST SCIENCE STUDIES:

Kourany, Janet A., ed. (2002). *The gender of science*, Prentice Hall, Upper Saddle River, N.J.

Lerman, Nina E., Oldenziel, Ruth, and Mohun, Arwen P., eds. (2003). *Gender & technology: A reader*, Johns Hopkins University Press, Baltimore, Md.

Mayberry, Maralee, Subramaniam, Banu, and Weasel, Lisa H., eds. (2001). *Feminist science studies: A new generation*, Routledge, New York.

Rothschild, Joan, ed. (1983). *Machina ex dea: Feminist perspectives on technology*, Pergamon Press, New York.

Index

237

About the Editor

MARGARET E. (PEGGY) LAYNE, P.E., joined Virginia Tech in 2003 as director of *AdvanceVT*, a program to increase the number and success of women faculty in the sciences and engineering. She previously served as a diversity consultant working with the American Association of Engineering Societies and as a fellow at the National Academy of Engineering, where she directed the program on diversity in the engineering workforce. Ms. Layne also spent a year as a Congressional Fellow sponsored by the American Society of Civil Engineers in the office of Senator Bob Graham (D-FL), where she was responsible for water, wastewater, and solid and hazardous waste policy issues. She has 17 years of environmental-engineering experience, and was formerly a principal at Harding Lawson Associates in Tallahassee, FL, where she managed the office and directed hazardous waste site investigation and cleanup projects. Ms. Layne has degrees in environmental engineering from Vanderbilt University and the University of North Carolina School of Public Health. She served as president of the Society of Women Engineers in 1996–97 and has chaired national committees for the American Society of Civil Engineers.